KV-421-292

MATHEMATICAL NEUROBIOLOGY

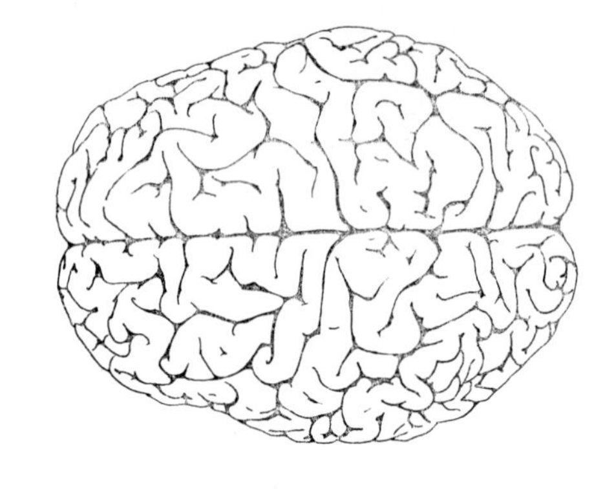

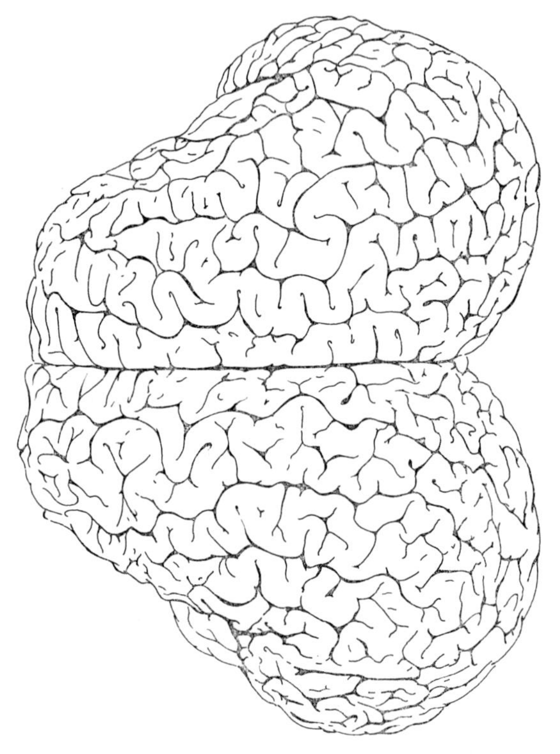

Frontispiece. The brain of normal adult man (top) and of the adult fin whale *Balaenoptera physalus* L. (bottom) seen from above on the same scale and with the front ends to the left. Both brains were formalin-fixed and had their outer coverings (dura and pia-arachnoid membranes) removed. Magnification ($\times \frac{1}{3}$), scale in centimeters. (Redrawn by R. Purcell from photographs by Tower, 1954).

MATHEMATICAL NEUROBIOLOGY

An Introduction to the Mathematics of the Nervous System

J. S. GRIFFITH

Professor of Chemistry and member of the Center for Neural Science at Indiana University, Bloomington, Indiana, U.S.A.

1971

ACADEMIC PRESS

London and New York

ACADEMIC PRESS INC. (LONDON) LTD
Berkeley Square House,
Berkeley Square,
London, W1X 6BA

U.S. Edition published by
ACADEMIC PRESS INC.
111 Fifth Avenue,
New York, New York 10003

Library of Congress Catalog Card Number: 71–141725
ISBN: 0–12–303050–1

PRINTED IN GREAT BRITAIN BY
THE WHITEFRIARS PRESS LIMITED
LONDON AND TONBRIDGE

Preface

This book is intended as a self-contained introduction to the study of the mathematical problems associated with the integrative action of the nervous system and especially of the brain. It tries to give a critical discussion of the relevant experimental facts and of various mathematical methods and techniques which have been used, with reference to further reading on each topic, and also to draw attention to some pieces of mathematical theory, such as the theory of non-linear equations and differential-difference equations, which are likely to be used more in future work on the brain then they have been yet. Unlike my previous book "A View of the Brain" it does not try to put forward a particular opinion about brain operation.

The book is based upon a course, which I have given here at Indiana University for the past two years, to graduate students coming mainly from a physical chemical or biochemical background, although I think it would also be suitable for physicists, mathematicians or theoretically-minded biologists. It is a pleasure to acknowledge how much I have learnt from discussion during this course and, especially about membrane properties, from L. Bass and W. J. Moore. I am also indebted to G. M. Wyburn for allowing me to use the four drawings of the human brain which appear in Chapter 1 and for permission to use other illustrations to J. C. Eccles, A. L. Hodgkin, G. Horn, U. Karlsson and D. B. Tower, also to Academic Press, the Johns Hopkins Press, Liverpool University Press and the Wistar Institute of Anatomy and Biology.

December, 1970

J. S. GRIFFITH
Department of Chemistry,
Indiana University

Contents

CHAPTER 1

Gross Features of Brains

1.1. Some Quantitative Data

Quantitative data about the nervous system, both at the gross and the detailed level, are much more sparse than is desirable and this certainly poses a major difficulty for attempts to build realistic mathematical theories of brain action. This is partly because the nervous system is an extremely complex structure containing, in the case of man, something of the order of 10^{10} nerve cells with perhaps 10^{14} interconnections, and partly because the collection of much of the data which would seem most fundamental to a mathematician, physicist or theoretical biologist, is difficult and tedious and has not appealed to the majority of biologists. Accordingly the first two chapters of this book, which attempt to give some quantitative background introduction, are inevitably very incomplete. In the present chapter much of the data comes from the compilation by Blinkov and Glezer (1968).

The human brain and spinal cord naturally occupy a unique position in our attention and four views of the brain are illustrated in Figs. 1.1 to 1.4 (from Wyburn, 1960). However, all vertebrate brains have a broadly similar structure and probably have similar principles of functional organization, at least in most respects. Invertebrate brains differ much more. It would not be appropriate here, even if I were capable of doing so, to attempt any survey of neuroanatomy and I shall merely refer to a suitable introductory book (Wyburn, 1960), and to some advanced ones (Cajal, 1952; Brodal, 1969; Truex and Carpenter, 1969).

Apart from processes which go to muscles or carry signals from sense organs, the vertebrate nervous system is practically completely contained within the rather sharply delimited region consisting of the brain and spinal cord. These latter two together are called the "central nervous system" or CNS. Some average measurements for man are given in Table 1.1. The predominant part of the brain mass is seen to lie in the cerebral hemispheres. This great development of the cerebral hemispheres is a very noticeable feature of primates and, among these, especially of man. It has often been believed to be connected with, or even to be the reason for, the apparently unique efficiency with which man can think abstractly and symbolically.

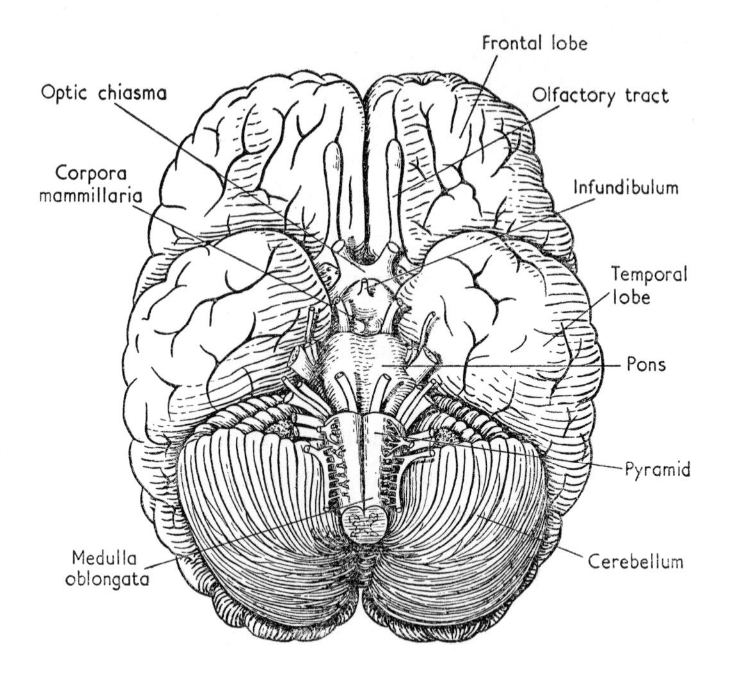

FIG. 1.1 The under surface of the human brain. (From Wyburn, 1960)

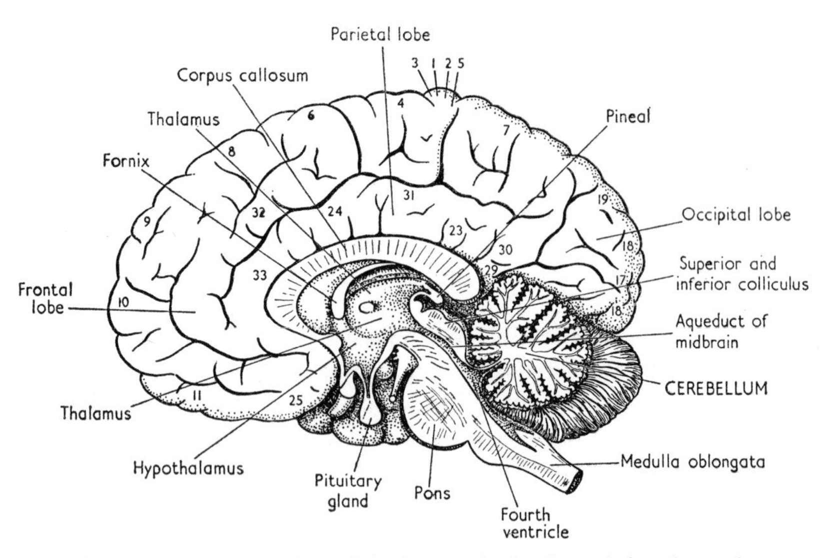

FIG. 1.2. Vertical cross-section of the human brain through its plane of approximate symmetry. In the upper and left-hand parts of the Figure we see part of the surface (cerebral cortex) of the right cerebral hemisphere. The numbers follow the Brodmann numbering system for regions of the cortex (see Ranson and Clark, 1961, Chapter 18). The primary visual receiving area is number 17, the major motor area is 4, while 1, 2, 3 and 5 are sensory areas. (From Wyburn, 1960)

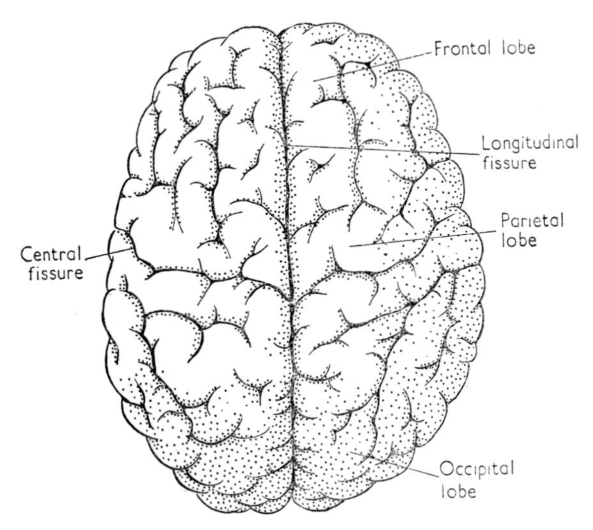

FIG. 1.3. The human brain seen from above. (From Wyburn, 1960)

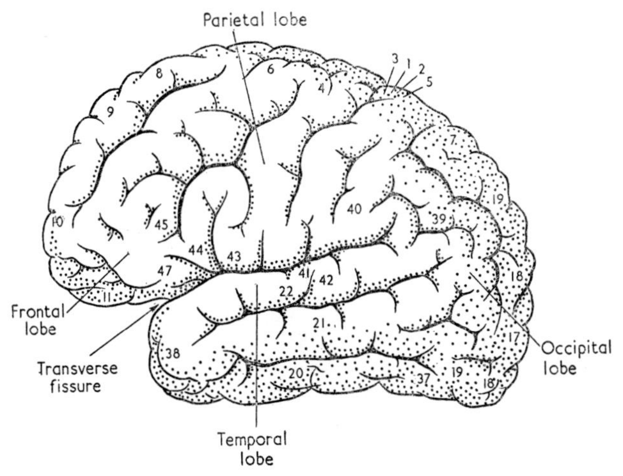

FIG. 1.4. The human brain from a side view. (From Wyburn, 1960)

TABLE 1.1

Some data for average human brains. (From Blinkov and Glezer, 1968)

			% of brain weight
Brain weight, male	1400 gm	Cerebral hemispheres	88
Brain weight, female	1300 gm	Cerebellum	10
Brain volume	1200 ml	Brain stem	2
Spinal cord weight	27–38 gm		
Spinal cord length	42 cm		

The development of the brain as a whole in relation to the spinal cord is also rather extreme in man, as is illustrated in Table 1.2, and this has also

TABLE 1.2

Spinal cord weight as percentage of brain weight. (From Blinkov and Glezer, 1968)

Animal	%
Man	2
Macaque monkey	12
Domestic cat	24
Tortoise	120

been considered to be a reason for man's intellectual predominance. Another thing which has been noted in this connection is the high degree of folding (called convolution) of the surface part (cerebral cortex) of the cerebral hemispheres, thus giving a relatively large surface to volume ratio, although man is not the most extreme animal in this respect (see Table 1.3). All these

TABLE 1.3

Areas of outer surface of cerebrum and of the cerebral cortex in cm^2. (From Elias and Schwartz, 1969)

Animal	Outer surface	Cerebral cortex	Ratio
Kangaroo	53·0	74·9	1.41
Fox	68	135	2.01
Man	795	2275	2.86
Bottlenose dolphin (Atlantic)	567	2700	4.47
Bottlenose dolphin (Pacific)	693	3343	4.75
False killer whale	1488	7392	4.97

arguments are extremely superficial however and, until we do understand in considerable detail how the human brain works, we cannot possibly tell whether similar things could be done by brains lacking particular gross features which happen to be always present in the brains of mentally normal humans.

With a digital computer one might well expect that, other things being equal, the larger it is or the more circuits or store that it has, the more things will it be able to do. Therefore it is natural to ask whether larger brains are generally better brains, and especially whether man's brain is the largest of the lot. The answer to the latter question is that man does have one of the largest brains but not the largest. A selection of average or typical brain weights is given in Table 1.4 and illustrates this point (also see the Frontispiece). There is, however, considerable variation in weight from one specimen to another, as is shown in Table 1.5. Over a large range of size there does not seem to be any clear-cut relation between intelligence and brain size, which is rather surprising.

TABLE 1.4

Brain weights in grams for various animals. (Crile and Quiring, 1940; Tower, 1954; Blinkov and Glezer, 1968)

Animal	Weight	Animal	Weight
Fin whale	6785	Domestic cat	25
Indian elephant	4400	Alligator	8·4
Porpoise (dolphin)	1735	Tortoise	0·3
Man	1400	Field mouse	0·2
Walrus	1126	Common toad	0·07
Orangutan	372	Cockroach	0·0002

TABLE 1.5

Weight of certain normal human brains in grams. (From Cobb, 1965)

Australian bushwoman	794	European man (average)	1400
Anatole France (at 80)	1017	Thackeray	1658
Japanese woman (average)	1250	Bismarck	1807
Walt Whitman	1282	Cuvier	1830
European woman (average)	1300	Daniel Webster	1895

It has also been suggested that the percentage of brain relative to the body is especially relevant to intelligence. There would seem to be much less reason to expect this and Table 1.6 shows some obstacles that this view faces.

TABLE 1.6

Percentage weight of brain to body. (From Crile and Quiring, 1940; Blinkov and Glezer, 1968)

Baboon	5·6	Cockroach	0·2
Porpoise (dolphin)	2·6	Elephant	0·2
House mouse	2·5	Gorilla	0·1
Man (18 years old)	2·4	Whale	0·005
Common squirrel	1·9	Brontosaurus	0·001

Finally, it could be argued that it is the number of nerve cells (or of connections), rather than the absolute mass of brain material, which is relevant. It is true that the number of nerve cells/ml is less, at least in the cortex, in the whale and elephant than in man, but the total number of cells is still probably larger (cf. Tower, 1954). In conclusion, although factors like large brain weight or high nerve cell content may well favor the development of intelligence, man does not seem to be uniquely distinguished by any such gross parameter (see Cobb (1965) and Holloway (1966) for more detailed discussion).

1.2. Constituents of the Central Nervous System

1. *Nerve cells, also called neurones.* These are, of course, the best known constituents. They are the rapidly conducting signalling elements with which we are mainly concerned in this book, and whose properties we shall describe in more detail in the next chapter. They are cells which are highly branched (see Fig. 2.1). The human CNS probably contains of the order of 10^{10} nerve cells. These cells do not normally divide in adult life, but they do die, as a result of which there is a gradual decline in numbers throughout life so that an old man may perhaps have only about a third as many nerve cells as he had when he was born. The arrangement of the nerve cells is highly complex and at least in part is specific and genetically determined, although it may well be partly random. There is much similarity in the structure of the nervous systems of all vertebrate animals.

2. *Glial cells, also called neuroglia or glia.* These may be said to fill in the spaces between the nerve cells and so to play a structural role. They also transport metabolites from the blood supply to the nerve cells and lead to the formation of scar tissue after brain damage. They are branched and are able to divide in the adult animal. The human brain contains about 10^{11} glial cells, which is probably more than the number of nerve cells, and their density in different parts varies, but not enormously (unlike the nerve cells, see Blinkov and Glezer, 1968, page 243). Note that, in many diagrams, nerve cell networks appear as rather extended structures with large spaces

in between. This is deceptive as the spaces are essentially completely filled with glia.

3. *Blood vessels.* The main arteries and veins lie outside the CNS with smaller branches penetrating inwards. They carry the blood which contains many nutrient and energy-giving materials, including glucose and oxygen.

4. *Cerebrospinal fluid.* The brain may be thought of as a very thick-walled hollow sack, whose walls are the nervous tissue. In the interior is a liquid, the cerebrospinal fluid, which is essentially blood filtered of its white and red corpuscles and containing very little protein (see Davson, 1967). The interior of the brain itself contains a number of interconnecting cavities, or ventricles, which are continuous with a narrow hole down the middle of the spinal cord. The cerebrospinal fluid is secreted in the ventricles and flows away down this hole.

CHAPTER 2

Nerve Cells and their Properties

2.1. General Remarks

The nerve cells are the building blocks of the signalling system of the brain. They are to it as the logic circuits, wires and elements of the magnetic core store are to a digital computer (though see Section 2.3). Accordingly in any constructive theory of the brain, we have to know the properties of the nerve cells and how they interact with one another. In this chapter we shall give a survey of these basic facts (for further introductory reading see Eccles, 1960; Roeder, 1963; Hodgkin, 1965; and Katz, 1966; and for another discussion of some numerical data, see Young, 1964). Before starting this, it is important to emphasize that nerve cells come in an almost endless variety of shapes, sizes, connections and excitabilities (see, for example, Bodian, 1962, 1967). It is possible to say very few things indeed and have much confidence that they are true of *all* nerve cells. Quite apart from this, our state of quantitative knowledge either of the relative or of the absolute numbers of cells having particular properties is not nearly as extensive as one would like. Especially is this so for those properties which determine the values of the various parameters which occur in the mathematical models which are discussed later. Thus we cannot talk with confidence of a "typical nerve cell" or even of a few kinds of typical nerve cell and, to be complete, would be forced to give an enormous catalogue of particular cases, for most of which virtually no data are yet available. However, there are a number of properties, such as excitability, development of an action potential, synaptic linkage, etc., which are generally considered to be characteristic of nerve cells and it is these which have normally been idealized in mathematical models. However, when reading the account later in this chapter and especially when considering the mathematical models used in the rest of the book, it should be remembered that the impression of uniformity of character which is often implicitly given for the cells is a vast oversimplification in most, if not all, cases, and that the numerical values which are used are based on very few measurements.

2.2. Properties of Nerve Cells

A nerve cell is a specialized cell with a nucleus and, although this is not experimentally proven, presumably with full inherited DNA content. However, nerve cells possess the peculiarity of not normally dividing during adult life.

2.2.1. Typical Mammalian Motoneurones

We shall now follow Eccles in describing a "typical" motoneurone of the spinal cord. A motoneurone is a neurone which signals muscle fibers to contract. As almost all output from the brain leads to activity in one or more muscles of the body, it follows that almost all commands from the brain

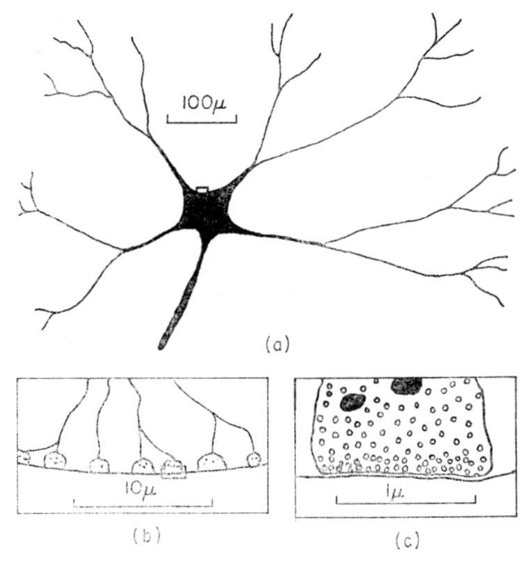

Fig. 2.1(a) to (c). Drawing of a motoneurone to illustrate general relationships of dendrites and axon to the soma. The small surface area that is outlined in (a) is drawn at 20 times higher magnification in (b) to illustrate the relationship of the synaptic knobs to the surface. The small area outlined in (b) is drawn at 10 times further magnification in (c) to show the width of the synaptic cleft and the thickness of the surface membranes of the synaptic knob and the nerve cell. Also shown are the synaptic vesicles and mitochondria of the synaptic knob. (From Eccles, 1960). Published with permission of the Johns Hopkins Press, Baltimore.

are finally executed through motoneurones. We must remember, however, that even motoneurones vary considerably one to another and that although many brain cells have similar properties to them, it is still uncertain whether the majority do.

A rather schematic representation of a motoneurone is shown in Fig. 2.1(a). It is composed of three parts. The black area in the center is the cell body, which contains the nucleus, and is often called the "soma" or "perykaryon" in the literature. In the example shown, it is about 70 μm across (1 μm = 10^4 Å).

The branched processes at the top are called dendrites. They may be 1 mm or more long. Finally, the process leading off from the bottom of the cell body is the axon. It gets surrounded by the so-called myelin sheath at some 50 μm to 100 μm from the cell body, which speeds up its signalling (see Section 2.2.5 and Hodgkin, 1965, Chapter 4). There is normally just one axon, which branches repeatedly further out. In the case of motoneurones, most branches go to muscle fibers but they can also terminate on other nerve cells. This latter mode of termination is normally the case for nerve cells other than motoneurones. An axon may be quite short or, for example in motoneurones controlling foot muscles, extend for up to a meter. The dendrites, together with the cell body and part of the initial piece of the axon give the input surfaces of the cell, whilst the output signal goes down the axon to its terminal branches, at a velocity typically in the range 1–100 meters/sec (see Tasaki, 1959; Buchthal and Rosenfalck, 1966).

2.2.2. Membrane Potential

The entire surface of the cell is bounded by a membrane, which is about 70 Å thick. The interior of the cell is negative with respect to the exterior, in the present case by about −70 mV. Although this may not seem a very large potential difference it leads to the high field strength of about 10^5 volts/cm across the very thin surface membrane.

TABLE 2.1

Concentrations of Na^+, K^+ and Cl^- in a motoneurone. (From Eccles, 1960) There is also some other unknown anion inside the cell

	mM External	mM Internal	Equilibrium potential (Nernst)
Na^+	150	~15	+60 mV
K^+	5·5	150	−90 mV
Cl^-	125	9	−70 mV

The membrane has a very complicated structure and its properties are fairly well understood at a phenomenological level but very incompletely at a molecular level. It is selectively permeable to ions, the most important of which are Na^+, K^+ and Cl^-. Their concentrations in the motoneurone are shown in Table 2.1. We do not have thermodynamic equilibrium here. This is illustrated by the entry in the last column of the Table which shows that internal potential difference V from the outside under which the given relative concentrations (strictly activities), c_e and c_i say, of each ion would be in equilibrium according to Nernst's equation

$$V = \pm \frac{RT}{F} \ln \frac{c_e}{c_i}, \tag{1}$$

where R is the gas constant, F the Faraday and T the absolute temperature (see Glasstone, 1946, p. 926; Hodgkin, 1965, p. 30). Evidently the chloride concentrations are very close to the equilibrium ratio but there is less sodium and more potassium inside than would be expected on that basis. This disequilibrium is maintained by a free-energy consuming process which is called the "active" transport of sodium and potassium ions through the membrane. There is, thus, not an equilibrium state but a steady state.

This steady state situation is called the resting state of the nerve cell and the potential difference of $V = -70$ mV which then exists between the inside and outside of the cell is called the resting membrane potential. On various assumptions V may be calculated from the equations of electrodiffusion of the ions through the membrane in terms of the concentrations on the two sides, the mobilities of the ions in the membrane, etc. For such derivations and further discussion see Planck (1890a, b), Goldman (1943), Schlögl (1954), Bass (1964, 1965), Guggenheim (1965), Bass and Moore (1967), Cole (1968).

Quite a good general understanding of the membrane potential may be obtained by using an equivalent electrical circuit, as discussed by Hodgkin and Huxley (1952). The rationale of this is that the disequilibrium of the sodium and potassium ions may be regarded as providing a pair of concentration cells in parallel with each other and having respective electromotive forces of $V_{Na} = +60$ mV and $V_K = -90$ mV, as given by the Nernst equation. The circuit is shown in Fig. 2.2, where g_{Na} and g_K are the conductivities for Na^+ and K^+ respectively and C is the capacity of the membrane. Other ions can contribute, as is shown by the presence of V_L in the Figure, but their effects can often be neglected and we shall do so here. For our typical motoneurone the total resistance of the membrane is about $R = 8 \times 10^5\ \Omega$ while $C = 3 \times 10^{-9}$ F, hence its time constant $RC = 2.4$ msec (see Eccles, 1960, p. 29).

Using this circuit we can deduce that the total current into the cell is given by

$$i = C\frac{\partial V}{\partial t} + g_{Na}(V - V_{Na}) + g_K(V - V_K), \tag{2}$$

where V is the membrane potential and we have neglected the effects of other ions. In the resting state, $i = \partial V/\partial t = 0$, so we can derive the expression

$$V = \frac{g_{Na} V_{Na} + g_K V_K}{g_{Na} + g_K} \tag{3}$$

for the resting membrane potential. This shows that V is a mean of V_{Na} and V_K, weighted according to the conductances. Thus $V_K < V < V_{Na}$. Actually in the resting state g_K is larger than g_{Na} and so V is nearer to V_K than to V_{Na}.

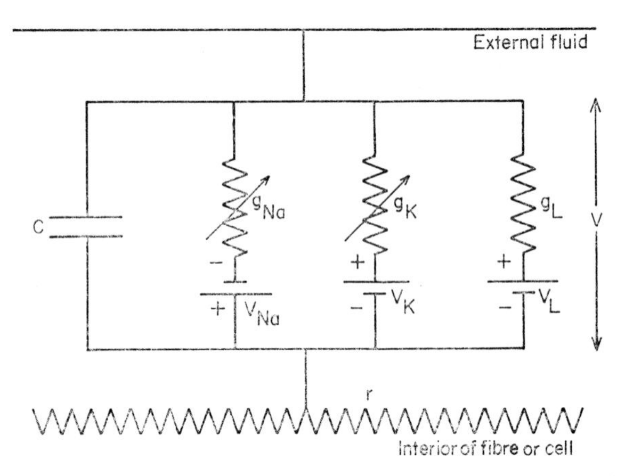

FIG. 2.2. Equivalent circuit for a patch of a nerve cell membrane. (After Hodgkin, 1965)

We shall use equation (3) in the next two sections and a word of warning is necessary. Equations (2) and (3) are useful as phenomenological equations in many cases. However, they do not have any adequate theoretical foundation and, indeed, the formula for V in equation (3) differs in its form from what one would expect on electrochemical grounds (see, e.g. Goldman, 1943). Hence they must be used with great caution even though they usually fit well qualitatively with experiment and are part of the quantitatively successful Hodgkin–Huxley theory.

2.2.3. Action Potential

The discussion in the preceding subsection relates to the whole cell and implicitly assumes that the cellular contents are perfectly conducting. This is not the case, which generally has to be taken into account in treating transient phenomena such as the conduction of the nervous impulse. The full treatment of nervous conduction is beyond the scope of this book, so we shall confine ourselves to a very brief introductory description, referring the reader elsewhere for a detailed account (see Hodgkin and Huxley, 1952; Fitzhugh, 1961; Hodgkin, 1965; Tasaki, 1967).

It is evident from equation (3) that the membrane potential could be changed by altering the ratio of g_{Na} to g_K. As we shall see shortly, this can happen under natural circumstances and so V can thereby change from its resting value of -70 mV. We now meet the feature of the membrane that leads to the possibility of nervous conduction, namely that the conductances

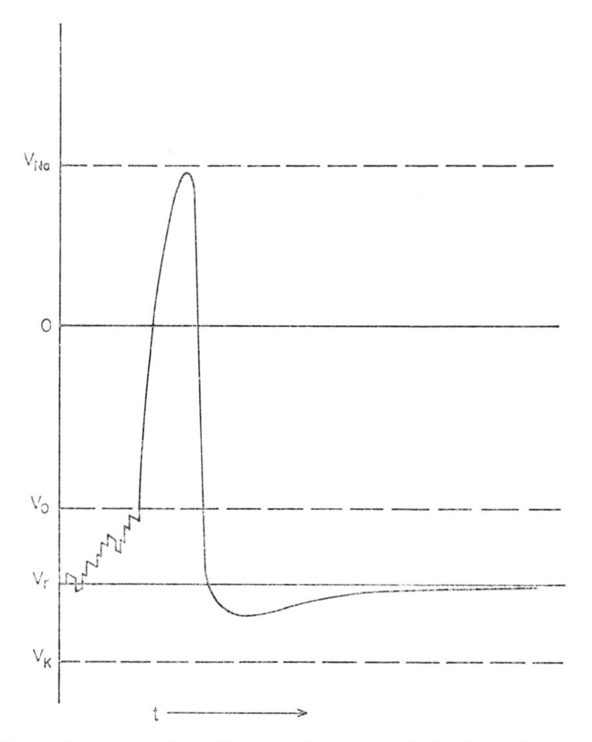

FIG. 2.3. The action potential. V_r = resting potential, V_0 = threshold potential. Duration of spike of the order of 1 msec; decay of hyperpolarization up to 100 msec. Before the spike, the influence of incoming EPSP's and IPSP's is shown.

g_{Na} and g_K are themselves dependent upon the potential difference V and in a rather peculiar manner at that, namely that they depend not merely on the instantaneous value of V but also on past values. Hodgkin and Huxley (1952) eliminate this dependence on past times in a very neat way, but only at the expense of introducing new variables whose physical significance is still rather obscure.

As a consequence of this dependence of g_{Na} and g_K on V, there exists in normal circumstances a definite threshold value V_0 (between about -60 mV and -50 mV) above which the mutual interdependence of g_{Na} and V leads to a runaway increase of g_{Na} relative to g_K with a consequent change of V within 1 msec or less to a value approaching V_{Na}. Because of the peculiar temporal dependence of g_K and g_{Na} on V, mentioned above, g_K increases also, but more slowly. Ultimately g_K again exceeds g_{Na}, pulling V back towards its resting value again. The observed potential changes associated with this sequence of events are called the action potential (illustrated in Fig. 2.3), in which the overshoot of the resting potential at the end is due to the relatively slow return of g_K from its increased value to its resting value.

The term depolarize is used to indicate an increase in V and hyperpolarize for a decrease in V. At the height of the action potential, the membrane is depolarized to an extent of about 100 mV relative to the resting potential. If one segment of the nerve cell membrane is depolarized in this way, it is easy to see that it induces a current flow and depolarization in the next segment too. Let the axon be arbitrarily, and approximately, split up into a sequence of "patches" each like that in Fig. 2.2 and connected by the resistances r, which are the resistances to current flow down the interior of the axon between successive patches. Then when, for example, g_{Na} is increased in one patch so that V across it rises from the resting value this causes a current flow through the next patch and depolarizes it. When this new depolarization reaches threshold, the g_{Na} of that patch has its runaway increase and so the action potential moves from patch to patch down the axon. This is the method of propagation of the action potential along axons and probably also along dendrites. It is clear even from this qualitative description that the rate of spread, i.e. the velocity of conduction, should be increased by lowering either the axonal resistance r or the membrane capacity C. Increasing the axonal diameter affects these two quantities in opposed directions as it decreases r but increases C. However, the net result actually increases the conduction velocity (see Katz, 1966, p. 92).

2.2.4. Postsynaptic Potentials

At each point at which the axon branches, the disturbance underlying the action potential, which is often called an impulse, proceeds along both branches. Finally, the axonal branches end in a number of minute enlargements, as we see in Fig. 2.1.(b). These are called synaptic knobs or synaptic boutons. An enlargement of one knob is shown in Fig. 2.1.(c). There is a gap of about 200 Å between the membrane of the knob and that of the cell on which it sits. The gap is called the synaptic cleft and contains fluid and possibly also macro-molecular bridges tying the knob to the surface of the other cell. The cell which has provided the knob at the synapse is referred to as the presynaptic cell while the other is the postsynaptic cell.

The synapses are the regions at which signals pass from one cell to another. Those shown in Fig. 2.1 are between a presynaptic nerve cell (or cells) and a postsynaptic motoneurone. The majority of the axonal branches from the motoneurone terminate in synapses with muscle fibers. These latter synapses are usually called neuromuscular junctions. When an action potential has passed down an axon and reached a synaptic knob, it normally stops there and does not jump immediately across to the postsynaptic cell. Rather, it stimulates the release of certain molecules into the synaptic cleft. These are contained within the synaptic knob in small packets, usually called vesicles, as shown in Fig. 2.1.(c), and it is generally believed that the action potential causes some of these vesicles to discharge their contents into the synaptic cleft.

The active material contained in the vesicles is called the chemical transmitter. Its effect on the postsynaptic membrane is locally to alter the conductances g_{Na} and g_K in equation (3) (probably also g_{Cl}) and thus alter V. Clearly, if the ratio $g_{Na} : g_K$ is increased, the membrane will be depolarized, while if it is decreased, the membrane will be hyperpolarized. Thus in the first case, V is moved towards the threshold V_0, and in the second case, away from it; whichever happens at a particular synapse depends both on the chemical nature of the transmitter and on the postsynaptic cell. A given cell will usually have many synaptic inputs and its internal potential V therefore continually moves up and down in response to the arrival of impulses at these synapses.

This motion is also illustrated in Fig. 2.3. Each arrival of an impulse at a synapse which causes postsynaptic depolarization leads to one of the short vertical rises at the left of the Figure. These are called excitatory postsynaptic potentials, usually abbreviated EPSP, because they shift the internal potential in the direction of V_0, which is the threshold value at which the action potential develops (excitation of the nerve cell). Impulses which cause hyperpolarization lead to a vertical fall shown in Fig. 2.3, which is called an inhibitory postsynaptic potential (IPSP). Because of the time constants associated with the capacity of the membrane, the EPSP and IPSP do not actually involve strictly vertical rises and falls.

The potential change due to an impulse arriving at a synapse does not persist indefinitely because there exist enzymes which destroy the transmitter. Thus a single EPSP or IPSP shows an approximately exponential decay towards the resting potential. The time constant of this decay arises from a combination of the kinetics of this enzymic destruction and of the electrical time constant mentioned above. It is typically of the order of a few msec (see Eccles, 1964, p. 42).

The chemical nature of the transmitters is not as well known as one would like. Acetylcholine is the best known (see Table 2.2 for formulae)

TABLE 2.2

Some important chemical transmitters

Transmitter	Formula
Acetylcholine	$CH_3COOCH_2CH_2N^+(CH_3)_3$
Norepinephrine ≡ noradrenaline	HO–[benzene ring]–$CHOH.CH_2.NH_2$, with OH on ring
GABA ≡ γ-amino butyric acid	$NH_3{}^+CH_2CH_2CH_2COO^-$

and is normally the excitatory transmitter at the vertebrate neuromuscular junction. It can also be inhibitory, as in its action on the vertebrate heart and in molluscs. Noradrenaline (not, as was once thought, adrenaline, whose formula differs by having $NHCH_3$ in place of NH_2) is another common transmitter and can have either an excitatory or an inhibitory action. GABA is an inhibitory transmitter in some invertebrates (e.g. crustaceans), and possibly elsewhere. For discussion, see McLennan (1963), pp. 82–86; Eccles (1964), chapters 5 and 12; Kravitz (1967); Hebb (1970).

Thus we have obtained a general picture of the mechanisms of movement of the internal potential of the postsynaptic cell under the influence of incoming impulses. When the potential V reaches the threshold value V_0, the cell develops its own action potential (we also call it a "spike" potential, or say the cell "fires"), as is also shown in Fig. 2.3. The effect of the various EPSP's and IPSP's on V is often, but not always, approximately additive. The addition of the effects from the various spatially separated inputs is often called "spatial summation" in the literature. It is also evident that impulses arriving within a millisecond or so of each other can add their effects to some extent, and this is called "temporal summation". Finally, it is experimentally observed and also a consequence of the Hodgkin–Huxley theory, that after a cell fires it cannot fire again for a short period (about 1 msec) and is relatively inexcitable for several msec. The first of these periods is called the "absolute refractory period", which will be important later, and the second is the "relative refractory period", which is usually ignored in mathematical models.

2.2.5. Miscellaneous Comments

1. *Classification of nerve cells.* A very good classification is given by the four kinds: motoneurones, neurosecretory cells (e.g. secreting hormones into the blood-stream, as in the pituitary gland shown in Fig. 1.2), sensory neurones and interneurones. The first two may be regarded as output cells

from the brain, the third as input cells to the brain (such as photoreceptor cells in the retina of the eye, or cells responsive to pressure on the skin) and the last one all the rest, i.e. all cells which have inputs only from, and outputs only to, other neurones.

More detailed classifications are based upon the shapes of cells and the arrangements of their axonal or dendritic processes (see Sholl, 1956, p. 45; Eccles, Ito and Szentagothai, 1967, p. 5) or on the nature of transmitter released (we say cells are cholinergic if they release acetylcholine, or adrenergic if they release noradrenaline). For discussion of the possible molecular basis of such classifications, see Griffith (1967a).

2. *Initiation of cell firing.* The initiation of the action potential in the post-synaptic cell, under the influence of the EPSP's and IPSP's due to the synaptic bombardment, is more complicated than we have indicated in our brief outline. The initial part of the axon has usually a lower threshold than the cell body itself and hence the action potential usually starts there both to travel down the axon and backwards to invade the cell body. For further detail, see Eccles (1960), pp. 47–56.

3. *Electrical synapses.* Although the arrival of an action potential at a synaptic knob normally has very little direct electrical effect on the post-synaptic cell, there are some synapses at which an appreciable postsynaptic potential is produced directly and there is no mediation of any chemical transmitter. Such synapses are called electrical synapses. For theory, see Eccles and Jaeger (1958) and for further discussion see Eccles (1964), chapters 9 and 14.

4. *Direction of propagation along axons.* Although axonal nervous conduction is normally outwards from the cell body, this is merely because the action potential is usually initiated there. The axon is equally capable of conducting in either direction, a fact which is often used experimentally in so-called antidromic stimulation. In this technique, an axon is artificially electrically stimulated at some point well out from the cell body, as a result of which an action potential travels up the axon to the cell body. Antidromic signalling also occurs naturally in spinal sensory nerves.

5. *Saltatory conduction.* The conduction along many axons is speeded up because they are wrapped in an insulating myelin sheath, mainly as a consequence of the resulting decrease of electrostatic capacity of the axon (see end of sub-section 2.2.3). They are called myelinated nerves. The sheath has gaps in it at intervals, called nodes of Ranvier, and the action potential jumps from one node to the next with an associated increase in conduction velocity. This mode of nervous conduction is called "saltatory". For anatomy see Wyburn (1960), pp. 5–7, and physiology see Hodgkin (1965), pp. 32–3 and chapter 4.

6. *Presynaptic inhibition.* It is often observed that axonal branches make

synapses with other axonal endings, just before the synaptic knob. It is believed that the function of this is, usually at least, to exercise a gating action over the latter endings, preventing a signal from passing to the synaptic knob if an action potential comes down the apposed "presynaptic" axonal branch. This is called presynaptic inhibition.

7. *Habituation.* Continued repetition of a stimulus to an animal often results in a gradual decrease of response. The exact mechanism of this is uncertain but, in part at any rate, it may be due to a gradual loss of responsiveness of neurones after repeated firing. Such a change could result from a rise in threshold of the cell (Salmoiraghi and Von Baumgarten, 1961). For further discussion, see Horn (1967).

8. *Dale's principle.* This asserts that the same chemical transmitter is used at every synaptic knob coming from one and the same cell (see Eccles, 1960, pp. 163, 184) and may well be fairly generally true (although see Wachtel and Kandel, 1967). It is intuitively quite plausible. However, there is less reason to expect the converse to be true, namely that all knobs synapsing onto a given cell should produce the same transmitter, and a counter-example to this is known in the mollusc *Aplysia* (Gerschenfeld, Ascher and Tauc, 1967).

9. *All-or-none character of neurones.* Nerve cell firing is often said to be all-or-none because either the threshold is reached, in which case a standard action potential proceeds down the axon with essentially the same amplitude and other characteristics every time, or it is not, in which case no signal goes down the axon at all. There is a certain analogy here with the digital character of processes in a digital computer, whose circuits are normally designed to deliver either a standard pulse or none at all. However, although the nervous system is evidently capable of complex timing activities (see particularly Horn, 1962; Kuiper and Leutscher-Hazelhoff, 1965), there is no reason to suspect that the whole mammalian brain is synchronized by a group of command cells in the way that much digital equipment has a synchronizing master oscillator.

10. *Non-linearity of output.* The effect on the internal potential V of the arrival of several presynaptic impulses at a cell is often approximately linear (i.e. additive). However, their effect on the output of the cell is not and is actually strongly non-linear. For example, if we have a cell whose threshold can be reached with EPSP's simultaneously from ten synaptic knobs, then if impulses arrive at five of these knobs, the cell will not fire so there is no output. If impulses arrive at the other five, there is still no output. But if they arrive at both of the five together, i.e. at all ten, the cell fires and gives its full output. The output from the sum of the two inputs is far from being the sum of the outputs which would arise separately from those inputs. This non-linearity is a very fundamental obstacle to mathematical theorizing

about the brain because, almost without exception, the mathematics of non-linear systems is vastly more difficult than that of linear systems of apparently comparable complexity.

11. *Uncertainty of output.* A piece of digital equipment is usually carefully designed to ensure that, on repetition of precisely the same input, the same output will appear. For example, even with random number generating programs on a computer, the same sequence of numbers is obtained a second time if the program is repeated with the same initial conditions. We shall discuss in Section 2.4 whether this is likely to be true for the brain but already mention here that the synaptic mechanism involves the release of a variable number of vesicles into the cleft and, as a consequence, when the postsynaptic cell receives a sufficient number of presynaptic impulses to bring it near to threshold, there must be some uncertainty as to whether it actually fires or not.

2.2.6. Some More Quantitative Data

As we remarked earlier, quantitative data about basic parameters are hard to obtain. In this section is collected some information, relating to man unless otherwise stated. We consider first the numbers of nerve cells in some of the sub-divisions shown in Figs. 1.1 and 1.2. These would be expected to vary from one individual to another.

1. *Cerebral hemispheres.* Most cells are in the cerebral cortex. The number of cells in the cerebral cortex (both sides of the brain) has been thought to be about $5–8 \times 10^9$. Recently Pakkenberg (1966) has argued that previous investigators have made insufficient allowance for shrinkage in the preparation of the brain, and finds a figure of 2.6×10^9 for a normal 18-years-old man.

2. *Cerebellum.* Probably about 10^{10} neurones (Blinkov and Glezer, 1968); it could be more if very small cells have been missed in microscopic examinations.

3. *Spinal cord.* 1.3×10^7 neurones (Blinkov and Glezer, 1968).

4. *Corpus Callosum.* This consists of the majority of the fibers which connect the two sides of the cerebral cortex, passing through the plane of approximate bilateral symmetry of the head and brain (see Fig. 1.2). It contains about 1.4×10^8 individual fibers (axons). See Blinkov and Glezer, 1968.

5. *Rods and cones.* The primary photoreceptors in an eye are the rods, which respond mainly to faint light, and the cones which respond to color in bright light. There are about $1.1–1.25 \times 10^8$ rods and 6.5×10^6 cones in each eye (Pirenne, 1967).

6. *Optic nerve.* Each eye gives off a bundle of axons going to the brain and which are called the optic nerves. It is still uncertain whether an optic

nerve contains any axons passing from the brain to the eye but, if it does, they must be a very small proportion of the total. Each human optic nerve has been found to contain 0.9 to 1.2×10^6 fibers (axons), with a mean of 1.01×10^6 for ten individuals (see Bruesch and Arey, 1942). The conduction velocity is mainly in the range 5–20 meters/sec (Sumitomo, Ide and Iwama, 1969, data for rat).

7. *Total input to brain and spinal cord.* There are 1.37×10^6 fibers into the spinal cord and, apart from the optic tract, 2.9×10^5 into the brain (Bruesch and Arey, 1942). These figures apparently refer to nerves on one side only and should be doubled to get the total, as the optic input is said to be 38%. The total therefore is about 5.3×10^6.

8. *Volume of cell bodies.* Hyden (1960) finds the following volumes in the rabbit (in μm^3, fixed tissues, except spinal neurones): cerebral cortex, 5×10^2 to 2×10^4; spinal neurones, 2.5×10^4 to 5×10^5; granule cells of cerebellum (one kind of cell found there, see Eccles, Ito and Szentagothai, 1967), 600–700; bipolar cells of retina (cells receiving impulses from the primary photoreceptor cells), 10^3 to 5×10^3.

9. *Number of dendritic branches.* Sholl (1956) finds 20–80 in cat cerebral cortex.

10. *Number of synaptic knobs on one postsynaptic cell.* A really detailed electron microscopic examination with precise pinpointing of each knob is available only for two cells of the lateral geniculate body of the rat (Karlsson, 1966). The lateral geniculate body is a relay station in the brain at which the axons of the optic nerve terminate, signals going on from its cells to the visual cortex. These two cells had respectively 83 and 133 knobs on the cell body and the nearest parts of the dendritic branches. Even in this work it was impossible to trace the dendrites to their terminations, so these figures may be underestimates. An interesting discovery by Karlsson was the presence on each of these two cells of a single third kind of process of unknown function, called a cilium, leading from the cell and apparently ending blindly between other cells without making any specialized contacts (see Fig. 2.4).

Indirect measurements by various authors indicate that it is very likely that cortical cells have many more synaptic knobs. Most recently Cragg (1967, 1968) has obtained figures of 6×10^3 to 6×10^4 per cortical cell in rat, mouse and monkey by dividing experimental estimates of number of synapses per unit volume by estimates of density of neurones per unit volume.

11. *Total number of synapses in human cerebral cortex.* This is not known, but if we assume Cragg's figures are not too far out for man also, we may combine them with Pakkenberg's work to obtain something between 1.6×10^{13} and 1.6×10^{14}.

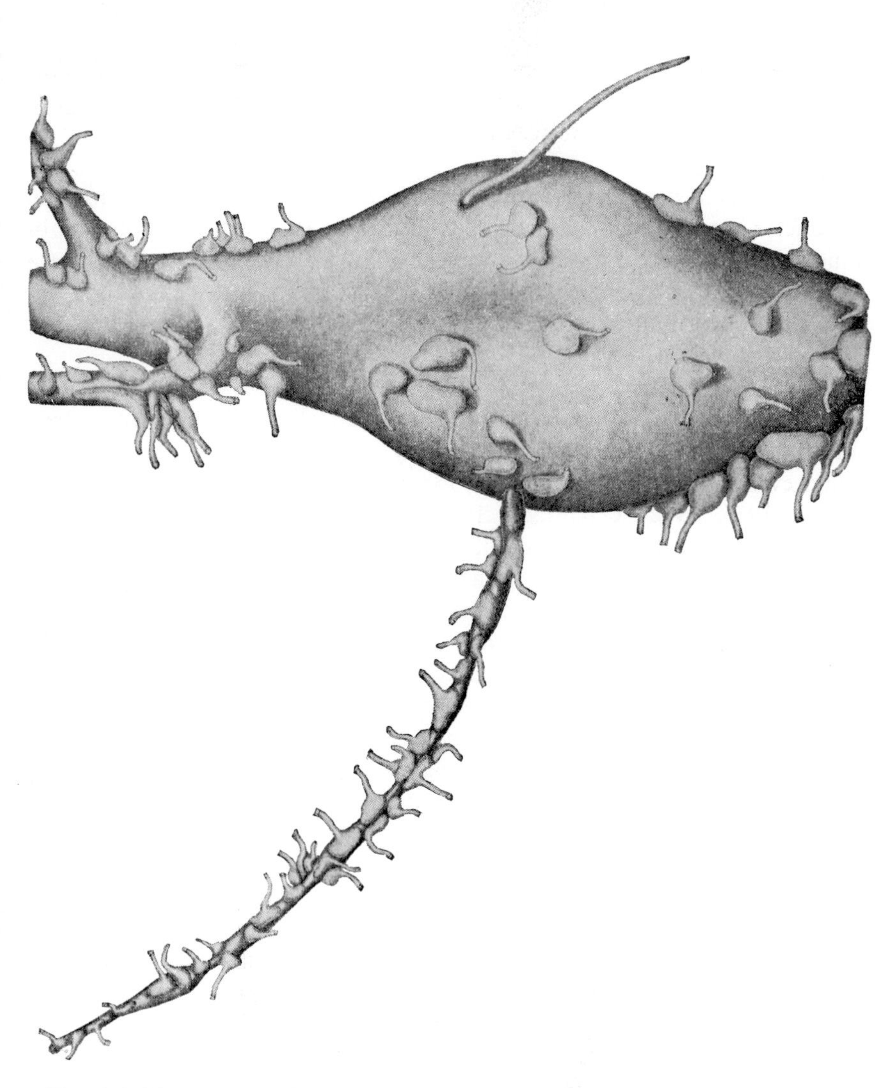

FIG. 2.4. Nerve cell body and nearest parts of dendrites of a neurone from the lateral geniculate body of the rat. Note the cilium at the top; the axon was not observed. (From Karlsson, 1966)

12. *Mean firing rate per cell.* This has been measured to be in the range 1–10 firings per second as an overall average in cat visual cortex by Herz, Creutzfeldt and Fuster (1964), Griffith and Horn (1966), Pettigrew, Nikara and Bishop (1968). It is difficult in these experiments to be sure that the cells whose firing has been recorded are typical, because one naturally tends to pick out the most active cells the most easily. Hence these rates may be overestimates.

13. *Threshold number of impulses needed to fire a cell.* This fundamental parameter is not known for brain cells. If θ is the number of synaptic knobs which need to receive impulses, then $\theta = 150–300$ for some spinal motoneurones (see Section 2.4.1), although θ must sometimes be much lower (see, e.g. Tapper and Mann, 1968).

2.3. The Glial Cells

For a long time, it has been speculated that glial cells may have some direct control over the signalling activity of the neurones and that they might even, at least in part, provide the physical basis of the memory trace (Glees, 1955, 1958; Galambos, 1961). One reason for considering this possibility is the large number of them for, as we have seen, there are probably more glial cells in the brain than nerve cells. The biological functions which have been assigned to the glia do not provide a satisfactory understanding of why there are so many. Another reason is that the glial cells occur in very close apposition to nerve cells, in some cases being so clearly related to a nerve cell as to be given the name satellite cell.

There is no *a priori* reason why the activity of the glial cells and of the nerve cells should not be coupled to some extent. Indeed, it would be rather surprising if they were not. The question at issue is whether such coupling is of significant functional importance for an understanding of the nervous activity of the brain, and this question must be regarded as still being a completely open one. However, Hydén (1962), has found some evidence of metabolic coupling between some brain neurones and glia and Svaetichin and collaborators have found a coupling between nervous activity, as measured by action potentials, and glial activity, as measured by intracellular potentials, in the eyes of some fishes. These latter workers have speculated that a similar coupling may occur generally in the brain and be highly significant to its function (Vallecalle and Svaetichin, 1961). Such an idea is very fascinating to consider, but is still within the field of speculation and it is probably unreasonable to have a strong opinion either way at present about the plausibility either of it or of the suggestion that glia are directly concerned with memory (see the discussion following the paper by Vallecalle and Svaetichin and also see papers in de Robertis and Carrea, 1965).

2.4. Variability of Output of Neurones

2.4.1. Quantization of Synaptic Transmission

The chemical transmitter is contained in vesicles in the presynaptic knob and, on the arrival of the action potential at that axonal ending, some of the vesicles are discharged into the synaptic cleft. This description contains the implication that an integral number of vesicles are discharged each time and that, therefore, the normal EPSP is built up from an integral number of contributions, or quanta. It is now generally believed that this is true, not only at the neuromuscular junction where this property was first discovered (Castillo and Katz, 1954), but also at all those central synapses which are mediated by chemical transmitters (see Katz, 1966; Martin, 1966), although some doubts have arisen very recently.

One effect of the quantization is to introduce a variability into the amplitude of the EPSP and we now discuss the theory of this. We start by making the assumption that all quanta are of the same amplitude and that the magnitude of the EPSP is proportional to the number of quanta from which it arises. As a consequence of the arrival of the action potential there is a certain probability, p say, that a given vesicle is discharged into the cleft. Let p be the same for all vesicles and let the probabilities be independent of each other. If there are n vesicles then the probability that exactly x are discharged (and that, therefore, exactly $(n-x)$ are *not* discharged) is given by

$$P_x = \binom{n}{x} p^x q^{n-x}, \tag{4}$$

where $q = 1-p$ and $\binom{n}{x}$ is the binomial coefficient $n!/(x!(n-x)!)$. With these assumptions, therefore, the EPSP varies in size on different occasions according to the binomial distribution (see, e.g. Cramér, 1955, Chapter 6).

The mean of the binomial distribution is $m = np$ and its standard deviation is

$$\sigma = \sqrt{npq} = \sqrt{mq}.$$

A useful measure of the variability in the number of vesicles discharged is given by the coefficient of variation, which is defined by

$$\text{C.V.} = \frac{\sigma}{m} = \left(\frac{q}{m}\right)^{\frac{1}{2}}. \tag{5}$$

As the EPSP is assumed to be proportional to the number discharged, it follows that its calculated coefficient of variation is also given by formula (5). In some experimental situations, it turns out that n is very large, while p is small in such a way that $m = np$ has a moderate value. Then the binomial distribution approximates to the Poisson distribution (see Cramér, 1955, p. 102):

$$P_x = \frac{m^x}{x!} e^{-m} \tag{6}$$

which has mean m, standard deviation $\sqrt{m}$ and C.V. = $1/\sqrt{m}$.

There are two ways in which these formulae need modification before they can be compared with experiment. The first is that the size of the EPSP is only expected to be approximately proportional to the number of vesicles released, as can be seen by considering formula (3). This non-linearity has been discussed by Martin (1966) and need not concern us here. The second arises from a variability in size of the individual quanta. Clearly the release of one single vesicle represents the minimum non-zero event which can occur at a synapse. The corresponding postsynaptic potential is called a "miniature" or "unit" potential and abbreviated to mepp. These miniature potentials may often be observed to occur spontaneously, i.e. even in the absence of the arrival of an action potential at the axonal ending. At neuromuscular junctions and some spinal motoneurones the amplitude of the mepp has been in the range 0.2–0.8 mV (Martin, 1955; Boyd and Martin, 1956; Kuno, 1964), but with a large variability in each individual case, presumably reflecting a variability in the amount of transmitter released per vesicle. When these two points are taken into account, the theory agrees well with the experimental data available, although the most thorough tests have been made at neuromuscular junctions at which the size of the EPSP has been artificially reduced (because of a reduction of p, rather than of the amplitude of the mepp) by bathing the junction in a solution high in Mg^{++} or low in Ca^{++}. Some values of m, n and p are given in Table 2.3, where n probably

TABLE 2.3

Estimates for parameters of quantal transmission at neuromuscular junctions. (From Boyd and Martin, 1956; Martin, 1966)

	n	m	p
Frog	700	100	0·14
Mammalian	700	300	0·45
Cat with high Mg^{++}	700	2·33	0·0033

does not correspond to the *total* number of vesicles in the synaptic knob, but may be the mean number close to the emitting membrane at one time (for review, see Martin, 1966).

We are now able to return to the question of the threshold number, θ, of synaptic knobs which need to receive impulses in order for the cell to fire. Kuno (1964) has made an experimental investigation of the input to spinal motoneurones, at which each unit potential is of the order of 0.2 mV and the threshold depolarization about 10 mV. He deduces a threshold to firing of 50–100 afferent axons (afferent means inward, efferent means outward) and finds the mean number of unit potentials per impulse per axon to be close to unity. Therefore the threshold number of vesicles needed to fire the cell is also 50–100. However, each axon probably branches at its end to give about three synaptic knobs so the value of m for a synaptic knob must be about $\frac{1}{3}$ and $\theta = 150$–300.

To what extent these numbers may be taken over to brain neurones is not clear. From one point of view one would say obviously not, they are quite different neurones. On the other hand, it would not be surprising if the two voltages on which the above numbers partly depend were not enormously different. One voltage is the threshold depolarization and the other the EPSP produced by the contents of one vesicle. However, this still leaves open the question of the "typical" value of m for brain neurones.

2.4.2. Reliability of Computers and Brains

The discussion in the preceding subsection shows that, on repetition of precisely the same input to a neurone, as given by an exact knowledge of the times of arrival of action potentials at each of its input synaptic knobs, the resultant EPSP will be variable. When the EPSP is near to threshold, a minor change in the number of vesicles released could make the difference between the cell firing or not firing. In this sense, there is an intrinsic variability or lack of determinism, in the input–output relations of a neurone (Griffith, 1967a, p. 43).

We cannot say at present how extensive this variability is for central neurones, nor how important it may be in brain function. However, we can ask about its possible significance. In digital computers and other digital equipment, the component circuits are normally deliberately built so that the input–output relations of their electric pulses are fixed and invariable. Any uncertainty in these relations which occurred in practice would be castigated as "unreliability". Indeed much theoretical work has been done on how to minimize the variability of the response of a complex piece of equipment in case it should have one or more unreliable components and the possible relevance to the brain has not been lost on these authors (see Von Neumann, 1956; Cowan and Winograd, 1963; Pierce, 1967; for introduction, see Arbib, 1965). But with digital equipment we normally wish to know where we are with it, at least to the extent of getting the same output for a given input repeated on a subsequent occasion. If this is not so, we talk irritably of its having "a mind of its own" and consider that it needs repairing. Of course there are devices, such as the British computer ERNIE which selects the Premium Bond winners (in a State lottery), which are designed to give an extremely variable output. Nevertheless, to date, those machines which have been loosely termed mechanical or electronic "brains" have also usually been those in which variability is frowned upon. When it comes to animal brains, however, these were "designed" by Nature and we do not know in advance whether she has our prejudices about reliability and variability. In fact it seems to me that there is a good theoretical reason, which we will discuss in the next subsection, for expecting her not to do so and thus for her to allow animal brains to have "minds of their own".

2.4.3. The Theory of Games

The term "the theory of games" is normally interpreted to refer to the classic work by Von Neumann and Morgenstern (1955) and to developments arising out of that work (for elementary introduction, see Vajda, 1966). It contains a great amount of elegant and satisfying mathematical theory and, parallel with this, a large number of concepts which were first explored at all clearly by them. Most of these have potential relevance in biology, but we shall concentrate here on one particular one.

Following Von Neumann and Morgenstern, we illustrate this concept by examining a game in which the issues involved appear in their simplest possible form. This is in the game of matching pennies, played by two players, A and B say. Each player, independently and unknown to the other, chooses either "heads" or "tails". They then compare their choices and if both have the same, i.e. both heads or both tails, one player, A say, has won. If they have chosen oppositely, i.e. one head and one tail, the other player B has won. A fixed sum of money is transferred, on each play, from the loser to the winner. The game is then repeated as many times as the players wish. There seems no reason to expect any one of the four combinations (heads, heads), (heads, tails) (tails, heads), or (tails, tails) to be favoured *a priori* over any other so, as A wins with two and B with the other two, the two players appear to be given equal opportunity in the game.

Nevertheless, as one will remember if one played this game as a child, a player, say A, may hope that by using his reason he will win more often than not. He argues that since B has to keep choosing either a head or a tail, it is likely in practice that there will be some pattern to his choices. For example, if B has just chosen heads for two plays in succession he may always feel that he should now switch to tails. If this is so, then A can predict this in advance and hence be sure of winning by choosing tails also. By finding out B's strategy, A can generally use the knowledge in order to win. The same opportunity exists, of course, for B.

Most players of this game soon realize that the way to avoid being taken advantage of in this manner is to try to make sure that one chooses heads and tails equally often but with no pattern to one's choices. Put more technically, the sequence of choices should be as if it was generated by a sequence of independent random variables, each referring to the two possible events "heads" and "tails" and having a probability of $\frac{1}{2}$ for each. This could be achieved by spinning a coin out of sight of one's opponent. It ensures that on the average one wins as often as one loses because, *irrespective* of what the opponent plays, one is equally likely to win or lose the fixed sum at *each* play. The mathematical expectation of gain is zero. On the other hand, providing one's opponent appreciates these considerations, one has no hope of doing any better than this because he, too, will choose heads and tails at random.

This exposes in a very pure form the fact that, in at least some competitive and recurring situations, there is no unique best way of proceeding but the best strategy may involve choosing quite randomly between two or more different ways, thus giving on different occasions a different overt response to what is apparently the same challenge. Von Neumann and Morgenstern show that it is the rule rather than the exception that, given a number of choices, the best strategy involves choosing each of these with a certain probability rather than choosing a certain optimum one each time. Animals are continually in competition with each other for things which are in short supply, such as food, space, mates, etc., and although their competition is not generally according to rules which would allow the existing theory of games to be applied without modification, it is certainly to be expected that the right sort of variability of response would give a selective advantage. We must therefore be prepared to contemplate the possibility that, in many brain neurones, the potential variability of output which is presumably there as a consequence of the mechanism of quantization of synaptic transmission (or possibly for other reasons) might be "deliberately" exaggerated to almost any extent under normal operating conditions. If so, then it would not be reasonable that such variability should necessarily be called "unreliability".

Before leaving this topic let us realize clearly that the theory of games is only ever remotely applicable to situations where there is genuine competition. Thus it is not applicable when an animal is dealing with an environment which, although presenting a variable and often undesirable aspect, is not actually adjusting its behavior adversely to that of the animal. For example in the laboratory, experiments have been performed in which an animal is presented with two levers, L_1 and L_2 say, and rewarded when it presses L_1, but not L_2, randomly on 70% of occasions and when it presses L_2, but not L_1, on the remaining 30%. These are called probability-learning tasks and it is clear that the animal's expectation of gain is greatest when it chooses L_1 every time. The theory of games does not enter in because the levers are not actively competing. The results of these experiments (for references see Russell, 1966) show that close to the best strategy is often adopted by the monkey or rat but more complex and less appropriate behavior is also often observed.

CHAPTER 3

Mathematical Models of Neurones

3.1. Logical Neurones

3.1.1. Introduction and Definition

Probably the best-known "mathematical neurone" is the logical or McCulloch-Pitts neurone (McCulloch and Pitts, 1943). It may be represented as shown in Fig. 3.1 and in its simplest possible form is a device which gives an output (to the right) if it gets an input from at least a certain number, say θ, of its inputs (on the left). It thus has a threshold θ which is, again in the simplest version of the model, a constant positive integer characteristic of the "neurone". θ is often written into the diagram of the neurone as in Fig. 3.1.

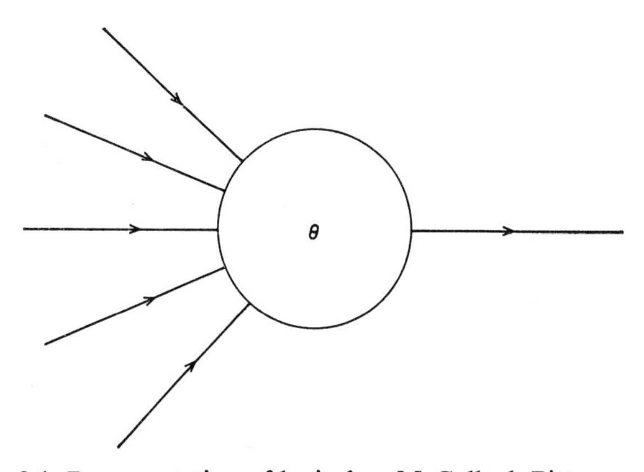

Fig. 3.1. Representation of logical or McCulloch-Pitts neurone.

The logical neurone purports to be an idealization of a real neurone and has the features of being able to be excited by its inputs (we include inhibition in a moment) and of giving an output when a threshold is exceeded. Its most peculiar feature is the way in which its behavior is a function of time. It is supposed that the neurone can only change its state at one of a discrete series of equally spaced times. Thus if one of these possible times is t_0 and the next is $t_0+\tau$, then the subsequent ones are $t_0+2\tau$, $t_0+3\tau$, ... The output from a given neurone at time $t = t_0+p\tau$ arrives as an input to all those to which it is linked, at the next time for change, i.e. at $t = t_0+(p+1)\tau$. A network of logical neurones behaves in a synchronized fashion, t_0 and τ being the same for every neurone in it.

Biologists often criticize the logical neurone for being too unrealistic, especially in relation to its time dependence. It is important for us to realize that this is rather unfair. The great advantage of the logical neurone is its simplicity, which often enables us easily to gain an insight into how a network of nerve cells might be expected to behave. It has many realistic features such as threshold, excitability, spatial summation and all-or-none output, but to gain simplicity we have to pay the price of lack of realism in some respects. This is a normal feature in the application of mathematics to the real world, namely that we often deliberately simplify in order to achieve mathematical tractability, whilst always recognizing the danger that such simplification can lead to spurious results.

After these introductory remarks, we now give a more formal abstract definition of a general logical neurone, under seven headings:

1. A logical neurone can exist in one of two states, which may be called "active" and "inactive".

2. It has one output, which can be connected simultaneously by one or more links to each of an arbitrary number of other logical neurones or to itself. This means it gives the same output along every link.

3. It has a total of n_e+n_i inputs, n_e of which are "excitatory" and n_i of which are "inhibitory". n_e and n_i can each take any non-negative integral value.

4. It has a threshold θ, which is normally taken to be a positive integer, although it could take other real values.

5. The neurone can only change its state at a discrete sequence of times $t = t_0+p\tau$, where p can take any integral value, and we often take $t_0 = 0$, $\tau = 1$. Each neurone keeps its state unchanged during each time interval $t_0+p\tau \leqslant t < t_0+(p+1)\tau$, where t_0 and τ are constants which are the same for every neurone of a given network. This is the assumption of quantized time, which is the most unpalatable feature of the logical neurone.

6. A particular input is active at time $t_0+(p+1)\tau$ if the neurone from which it comes was active at time $t_0+p\tau$ (we shall also then say that that neurone fired at time $t_0+p\tau$). We write N_e for the number of excitatory inputs which are active and N_i for the number of inhibitory ones which are. Evidently $N_e \leqslant n_e$ and $N_i \leqslant n_i$. N_e and N_i are, of course, functions of time and this may be made explicit if necessary.

7. A neurone is active at time $t_0+(p+1)\tau$ if and only if $N_e-\phi N_i \geqslant \theta$ at that time. ϕ is a positive real number characteristic of the neurone and will usually be taken to be an integer. Like θ, n_e and n_i, ϕ may differ from

one neurone to another, although it will often be interesting to investigate networks in which they have the same values for each neurone. Note that a more general rule of the type $\phi_1 N_e - \phi_2 N_i \geqslant \theta_1$ can be got into the simpler form given above by dividing through by ϕ_1, and writing $\phi = \phi_2/\phi_1$, $\theta = \theta_1/\phi_1$.

A logical neurone is a binary device, because it has two possible states. It is often convenient to represent its state in binary arithmetic notation, saying it is in the state 0 when it is inactive and in the state 1 when active. If we have a network of n neurones we can then number those neurones from 1 to n and represent the state of the whole network at a given time by a binary integer. Thus if a network contains just three neurones, the binary integer 101 signifies that neurones one and three are active, while neurone two is inactive. Evidently, at any given time, a network of n neurones has 2^n possible states.

3.1.2. Examples

We now consider a few simple illustrative examples, using the symbol $\rightarrow$ for an excitatory input and $\nrightarrow$ for an inhibitory one.

1. One neurone having $n_e = 2$, $n_i = 0$, $\theta = 1$. Because $n_i = 0$ it is unnecessary to specify ϕ. The diagram is

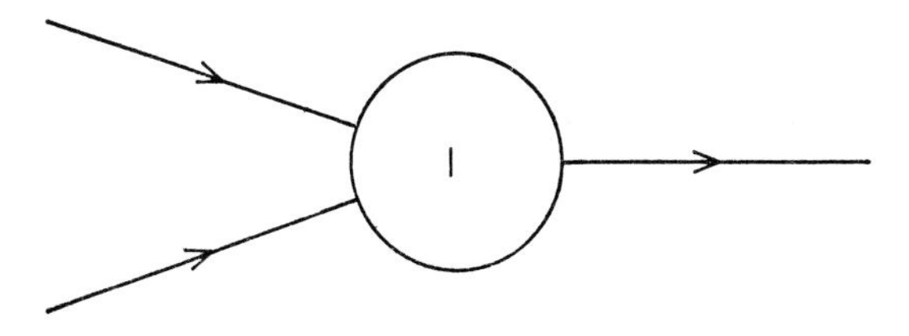

and the neurone fires if $N_e \geqslant 1$, i.e. if one input or the other or both are active.

It is interesting to note that, had we set $\theta = 0$, we should have got an output even if there were no input. Such a neurone could be called "spontaneously" active, although this would differ somewhat from the more common usage of the word "spontaneous" given in Section 4.1.

2. One neurone having $n_e = n_i = 1$, $\theta = \phi = 1$. The diagram is

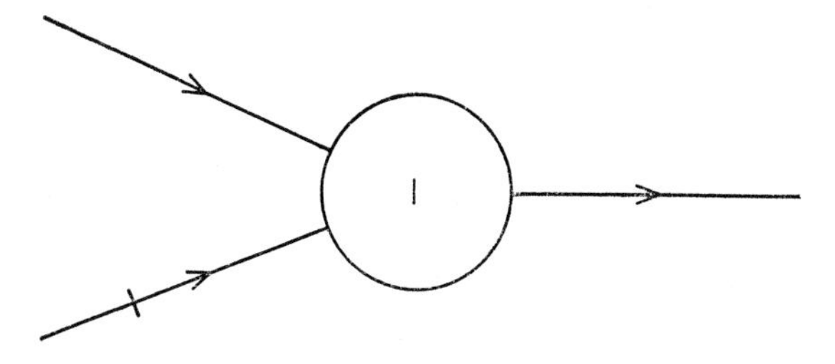

There are four possible input situations which can be tabulated most conveniently as follows:

N_e	N_i	$N_e - \phi N_i$
1	1	0
1	0	1
0	1	−1
0	0	0

We see that the threshold condition $N_e - \phi N_i \geqslant \theta$ is only satisfied in one case, namely when the excitatory input is active but the inhibitory one is inactive. It also follows from the table that, if we set $\theta = 0$, the neurone is spontaneously active but could be turned off if $N_e = 0$ and $N_i = 1$.

3. Three neurones, all excitatory links, i.e. all $n_i = 0$.

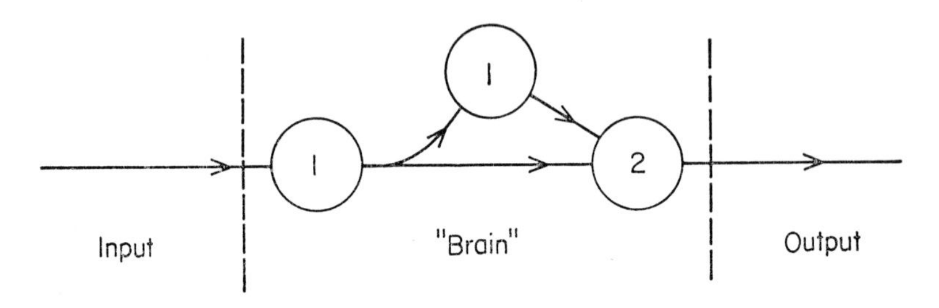

It is easy to see that we only get an output to the right after there have been two inputs successively on the left. One may thus think of the network as a very primitive "brain" which only reacts to repeated stimuli, but not to temporally isolated ones.

4. Self-re-exciting systems.

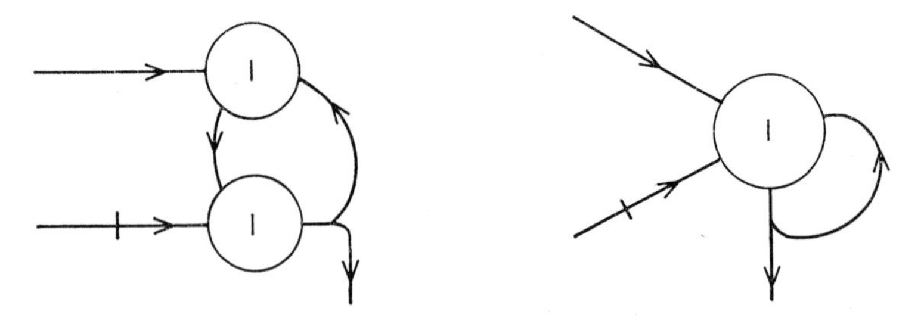

In each of these cases a single input at the top left corner continues to go round and round for ever unless it is "erased" by sending in an inhibitory input from the bottom left-hand side. The output from such systems can be changed indefinitely by a single input at one time, which may be far in the past. They could thus serve as primitive memories. It has been suggested (see Section 6.5.1) that human and animal memory might be based on this self-re-exciting potentiality of neural networks but it is generally considered that this is unlikely for long-term memory.

5. Another network of three neurones.

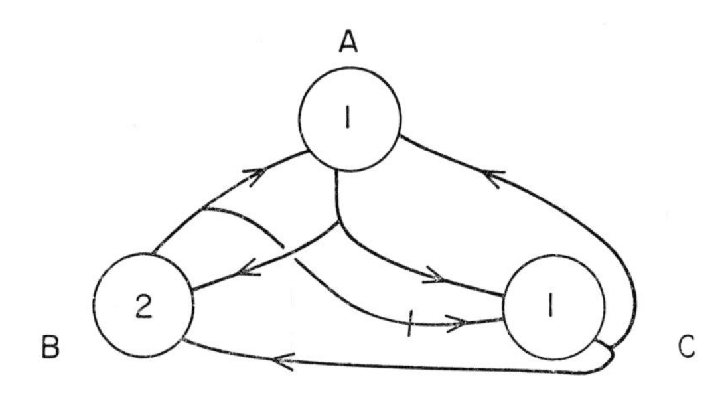

We now consider an isolated network of three neurones and ask what happens subsequently if the network starts in a particular state. Treating this as a purely mathematical problem, it is unnecessary to ask how the network is put into its initial state. However, it is evident that any initial state could be achieved by suitable inputs from the outside to the three cells.

It is convenient to use binary notation for states of individual neurones and a vector notation for a state S of the network, thus $S =$ (state of A, state of B, state of C). There are eight $(= 2 \times 2 \times 2)$ possible states S and we shall have completely characterized the behavior of the network when we have tried each of these as initial states. We then get the following diagram

$$(0, 1, 1) \longrightarrow (1, 0, 0) \rightleftarrows (0, 0, 1)$$
$$(0, 1, 0) \nearrow \qquad \nwarrow (1, 1, 0) \longleftarrow (1, 1, 1) \longleftarrow (1, 0, 1)$$
$$(0, 0, 0) \longrightarrow (0, 0, 0)$$

which the reader may easily verify. It shows that if we start the network in any state except the completely inactive one, it settles down to an oscillatory activity in which B is inactive but excitation shuttles between A and C.

This concludes our examples which were given to clarify the nature of the logical neurone. It should be emphasized that in a real brain we are concerned with so many cells (up to 10^{10}), each having perhaps 10^4 or more inputs, that we cannot expect to analyze its activity in the detailed way that we have in example 5 above. The sort of approach we must adopt then will be discussed in Chapters 5 and 8.

3.1.3. Connection with Real Neurones

We have remarked already that many features of real neurones are well represented in the logical neurone, but that the quantization of time is not realistic. Nevertheless, we should like to have some idea of what value and significance to assign to the time interval τ. In my opinion, τ should be regarded as an average estimate of the time taken between the firing of one presynaptic cell and the time when the effect of that firing first has the potential of influencing the firing of those postsynaptic cells to which the first cell is linked. Then τ is made up of four components, each of which would probably be a small multiple or submultiple of 1 msec. The first, τ_1, is the time for the action potential to travel down the axon to the synaptic knobs (if the velocity of conduction is v meters/sec and the axonal length is l mm, then $\tau_1 = l/v$ msec). There is then a slight delay τ_2, termed the synaptic delay, before the postsynaptic potential (PSP) starts to appear (Eccles, 1964, p. 42 gives some values of τ_2 as 0.2–2 msec). The PSP then appears, giving τ_3. Whether τ_3 should be the time to the peak of the PSP or until it has, say, half decayed from its peak, is a little unclear; probably the latter. Anyway τ_3 again is probably usually a few msec (see Eccles *loc. cit.*). Finally, τ_4 is the time of rise of an action potential from the threshold to its peak. τ_4 is probably typically less than 1 msec. Thus we should think of $\tau = \tau_1+\tau_2+\tau_3+\tau_4$ as being a few msec, probably usually less than 10 msec.

Although this gives a way of assigning a value to τ, it does not make the quantization acceptable. It requires that we should only allow a cell to fire at the times $t_0+p\tau$. The arbitrariness of the choice of t_0 is immediately apparent: since there is normally no synchronization in the real system, there is no reason to prefer any one value of t_0 over another.

We shall not pursue this undoubted defect of the logical neurone any further here, but merely remind the reader that these "neurones" are useful in theoretical discussions because of their relative simplicity, and mention two other points. The first is that we have not explicitly considered the question of refractoriness (Section 2.2.4). The refractory period would normally be less than τ and so the very fact that a logical neurone cannot fire twice in less than τ seconds is sufficient to deal with it. Were that not the case, one could introduce into the definition of the logical neurone the requirement that it could not fire again until at least $x\tau$ seconds after it last did so, for some fixed integer $x > 1$. The second is that habituation can easily be incorporated into the definition by imposing some restriction on the number of times a logical neurone can fire in a given period. For example, one could require that it cannot fire at time $t_0+p\tau$ if it has fired more than a times in the time interval $t_0+(p-b)\tau \leqslant t < t_0+p\tau$, where a and b are fixed integers (note that the refractory period is given by the special case $a = 0$, $b = x-1$).

3.2. Real Time Neurones

3.2.1. Basic Definition

From a functional point of view we know a very great deal about the activity of a neurone when we have a record of its internal electrostatic potential and of how this has been altered by EPSP's and IPSP's due to activity in attached cells. This suggests that we may define a much more realistic mathematical neurone by concentrating attention on this internal potential as a measure of the state of a neurone, which is thus characterized by a parameter V (Gluss, 1967; Griffith, 1967a). It is most convenient to choose V so that its physical significance would be the deviation of the internal potential from its resting value of around -70 mV. So the resting potential corresponds to $V = 0$.

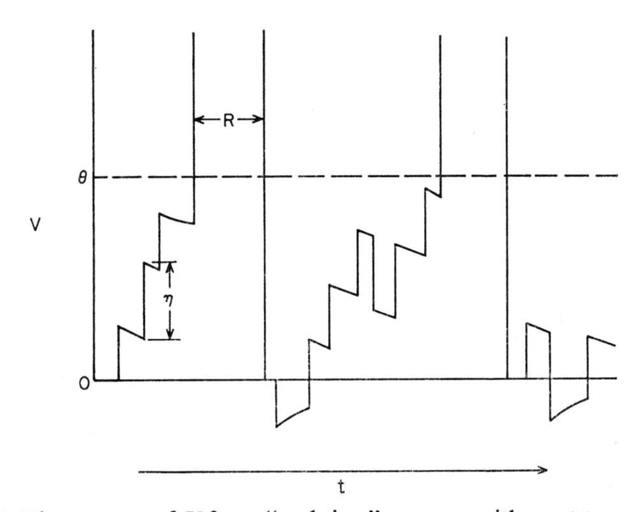

Fig. 3.2. Time course of V for a "real time" neurone with reset to zero assumption. (Compare with Fig. 2.3)

Three things can happen to V (see Fig. 3.2 for illustration).

1. If a presynaptically attached, i.e. input, cell fires at time t, then at time $t+\delta$ we alter V to $V+\eta$. δ represents the delay between the time the input cell reaches its threshold to firing and the peak of the resulting EPSP or IPSP. δ is the same as τ as defined in Section 3.1.3, providing τ_3 is taken as some-

where between zero and the time to the peak of the PSP. η corresponds to the height of the EPSP, in which case it is positive, or the IPSP, in which case it is negative.

2. At any time t' at which V changes from being $< \theta$ to being $\geqslant \theta$, we say the cell fires. θ is called the threshold. For $t' < t < t'+R$, we set $V(t) = \infty$ (or some large value $> \theta$). We then put $V(t'+R) = 0$. R is called the refractory period (it corresponds to the absolute refractory period of a real neurone). Both R and θ are fixed constants characteristic of the cell.

3. At all times not covered by rules 1 and 2, V satisfies the differential equation

$$\frac{dV}{dt} = -\varepsilon V. \tag{1}$$

ε is a fixed positive constant, characteristic of the cell, and corresponds to an average estimate of the time constant of decay of isolated EPSP's or IPSP's. ε^{-1} is probably of the order of a few msec.

3.2.2. Remarks

There is no reason why θ, R and ε should not differ from one neurone to another nor why δ and η should not alter from one synapse to another.

Although it is natural to interpret V as being related to the internal potential, there is another interpretation which might be more correct for some neurones. We remarked earlier (Section 2.2.4) that the time course of a PSP is a combination of the effects of the discharge of the membrane capacity and of the enzymatic destruction of the transmitter. In the limiting case of a very short electrical time constant and long life of transmitter, the PSP would strictly follow the transmitter concentration on the post-synaptic membrane. V could then be interpreted as "total transmitter level" at all input synapses to the cell, counting transmitter at excitatory synapses as "positive" and at inhibitory ones as "negative". Since there is no reason to expect the internal potential to follow the transmitter level during an action potential, this would mean that we could consider replacing rule 2 above with the following extreme alternative: at any time t' at which $V(t') \geqslant \theta$ we say the cell fires, providing it has not fired previously in the period $t'-R < t < t'$. In that case $V(t)$ would satisfy the equation (1) at all times not covered by rule 1 alone.

We thus have two possible limiting versions of the model, one in which V is reset to zero at time R after the cell fires and the other in which V is unaffected by the actual firing of the cell. It is probable that the first is usually nearer to the truth (see Eccles, 1964, Section 4B) and in its strict form implies that the value of V after a firing is totally unaffected by the extent to which V exceeds the threshold θ at the time of firing. This is obviously not so with the second version, with which it is obviously possible in principle for V to become so large that, in the absence of further input, it would cause the cell to fire twice or more successively. Using equation (1), this would happen if V were suddenly raised to a value V_0 satisfying $V_0 \geqslant \theta\, e^{\varepsilon R}$. It is probable that the repetitive response of certain cells in the spinal cord (Renshaw cells, see Eccles, 1964, pp. 119–121) is largely due to this kind of mechanism.

Habituation as an effect on the threshold of a cell may be conveniently introduced by allowing θ to vary with time according to the following rules:

1. When the cell fires alter θ to $\theta+h$.
2. At all other times θ satisfies

$$\frac{d\theta}{dt} = -\beta(\theta-\theta_0). \tag{2}$$

h, β and θ_0 are positive constants characteristic of the cell.

3.3. Computer Simulations

3.3.1. Logical Neurones

Many workers have simulated neural networks on a digital computer (see, e.g. Farley and Clark, 1961, and Harmon and Lewis, 1966). We shall not discuss detailed programming problems here but draw attention to a few numerical points relating to the space and time requirements for such simulations.

With a network of n neurones we have the following space requirements:

1. *Values of θ and ϕ for each neurone.* For a general network we need to store $2n$ numbers but, if we assume all neurones have the same values for these parameters, we need only store 2.

2. *Connectivity of the network.* For each neurone we must say how many links it has to each other neurone and whether they are excitatory or inhibitory (by putting + or − in front of the number). This requires up to n^2 numbers. Alternatively we could give a rule to determine whether neurone x is linked to neurone y or not. For example, we could say that if the remainder after dividing xy by n lies between 1 and 10 we have one excitatory link from neurone x to neurone y but otherwise there are no links. Such rules are easy to program, require a space in the store which is essentially independent of n, and can be useful in setting up pseudo-random networks to simulate the random networks discussed in Chapter 5.

3. *Present and immediately preceding state of network.* The present state is always calculated from the preceding state, and the latter must generally be stored until the present one has been completed. This means storing $2n$ numbers, each being 0 or 1.

Thus as far as space requirements are concerned, a computer can simulate a network containing about as many neurones as it can store numbers. In other words, even 10^{10} is not out of range providing the connectivity is largely specified by giving rules rather than actually enumerating all the links (which for the human brain would still only need 10^{14} or so numbers, rather than $n^2 = 10^{20}$).

In considering time requirements, the important time is the time T required to calculate the new state of the network from the preceding one. If each neurone is linked to q others then this is approximately given by $T \approx n(qt_1 + t_2)$, where t_1 is the time required to calculate the contribution of a given link to the sum $N_e - \phi N_i$ and t_2 is the time to see whether the resulting sum is $\geqslant \theta$. As a consequence, the time requirements impose a much more serious restriction on the size of network which can be examined than do the space requirements. For example, if $qt_1 + t_2 = 100$ μsec and $n = 10^4$, 10^6 or 10^{10}, $T = 1$ sec, 1 min 40 sec or $11\frac{1}{2}$ days respectively.

Thus, even if we knew all the necessary parameters, putting a McCulloch-Pitts version of the human brain on a computer would pose more of a problem of speed than of storage space. In view of the great speed of a modern computer, this must appear something of a paradox. In fact it is resolved by realizing that the normal version of digital computer is ill-designed for handling this sort of simulation because its central arithmetic unit can only operate at one time, albeit very fast, on a very small amount of the data in its store. If we knew enough to simulate the human brain, we would use separate electronic circuits for each neurone (quite realistic analog circuits even for neurones operating continuously in time have been described, see Harmon, 1959, 1961) and would have these circuits working simultaneously in parallel. Thus the time problem would be removed and, in fact, the artificial brain could be built to operate faster than the real one because the electronic time constants could certainly be in the microsecond and probably the nanosecond range (if memory could be suitably incorporated). If the latter were achieved, such an artifact could get through 100 years ($= 3 \times 10^9$ sec) of human thought in about 50 minutes.

3.3.2. Real Time Neurones

With a digital computer simulation, under each neurone at time t we need the current value of V (and of θ if it is allowed to vary) and also the times of firing of all attached presynaptic neurones back to the times $t - \delta$. By storing and continually updating this information we calculate the evolution of activity in the network as a function of time. Two points may be made about programming such a calculation. First, because a cell can only fire when an impulse arrives from another cell (i.e. on an upward jump of η, see Fig. 3.2), we need only calculate V at such times (which are known because it is known when the input cell has fired). Hence to advance the computer simulation of the network by a given finite time δt we need only perform a finite number of calculations, even though V is a function of the continuous variable t. Secondly, although it may seem necessary to calculate continually which cell fires next in the whole network, this is not so because no cell can influence another until at least δ_{min} (the smallest of the delays δ) after it has fired. Hence we can most conveniently perform the calculation by going through the neurones in order and calculating for each if and when it fires in the next time interval δ_{min}. This simplifies the programming problem considerably, and was used in my previous work (Griffith, 1967a, Chapters 4 and 5).

Finally, the time and space requirements are considerably increased, depending on circumstances, by a factor of ten at the very least or probably much more.

3.4. Symbolic Logic and Switching Circuits

3.4.1. Symbolic Logic

McCulloch and Pitts (1943) pointed out an interesting isomorphism between the input–output relations of their idealized neurones and the truth functions of symbolic logic. This has attracted a lot of attention and many people have thought that it casts great light upon brain function and the neural basis of the logic of human thought. Personally I do not believe that this is so, at least to date, nor do I think that the logical notation which thus becomes available to describe neural activity has much real use. I think the latter because the logical expressions required to describe a neurone seem to me much more cumbersome and difficult to manipulate than other more usual ones, especially when any large number of neurones or interconnections are being considered, and because of the difficulty (Kleene, 1956) of dealing with networks which, like examples 4 and 5 of Section 3.1.2, have any re-entrant paths (which probably includes all networks of any biological interest). Not everyone would agree with this verdict and so we give here a brief introduction to the matter to help the reader form his own opinion (for a clear elementary account of symbolic logic, see Basson and O'Connor, 1965).

The relevant part of symbolic logic is concerned with the question of the truth of composite statements, given the truth or falsity of the constituent simpler ones. In elementary logic there are only two alternatives considered: everything is either true or false. Composite statements are formed from simpler ones by combining them, using certain logical symbols. We now introduce some of these, using symbols like x or y to stand for the simple statements.

1. Or. The symbol is V and from x and y we can construct xVy (also yVx which is identically the same). The definition of a logical symbol is exhibited by a so-called truth table which shows whether xVy is true or false when you know whether each of x and y separately are. The defining truth table in this case is:

x	y	xVy
True	True	True
True	False	True
False	True	True
False	False	False

Thus, in ordinary English, the logical "or" means "either one or the other or both".

2. And. The symbol is a dot "." and from x and y we construct $x.y$. Again we can write a truth table. This time we shall note the binary character of the alternative "true" or "false" and write a "1' to indicate "true" and a "0" to indicate false, obtaining the table:

x	y	$x.y$
1	1	1
1	0	0
0	1	0
0	0	0

3. "Not" is written $\sim$, and logical implication (i.e. if this, then that) is written $\supset$. The truth tables are:

x	$\sim x$
1	0
0	1

x	y	$x \supset y$
1	1	1
1	0	0
0	1	1
0	0	1

$x \supset y$ is identically the same as $(\sim x)Vy$, i.e. if the first of these statements is true so is the second and vice-versa, and as the reader will readily see on examining its truth table does not have quite the same meaning as implication does in ordinary English (we do not normally say that a false proposition implies the truth of a true proposition). However, this is the way in which logicians have found it useful to define it.

McCulloch and Pitts now draw attention to the fact that if we let "0" in the truth function correspond to "not firing" and "1" to "firing", then the truth functions are very like certain simple McCulloch-Pitts neurones. The corresponding neurones are:

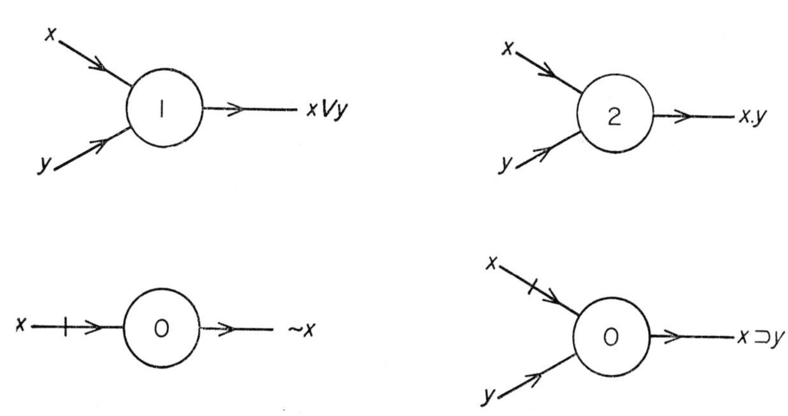

The truth tables we have given are also the tables showing the input–output relations of these neurones. Conversely, any neurone can be represented by a truth function. Suppose the neurone is

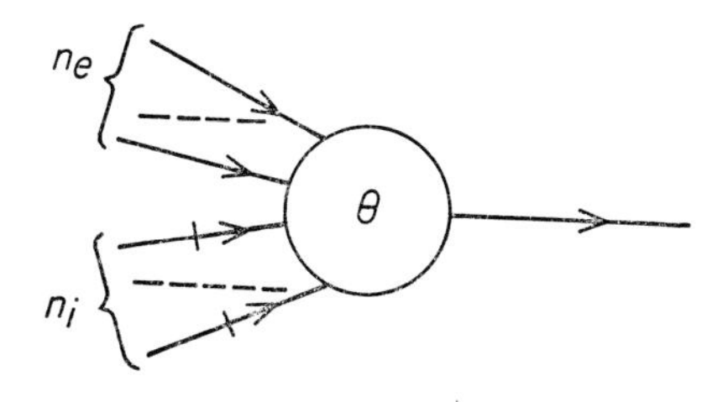

having n_e excitatory inputs which we represent by $x_1, x_2, \ldots, x_{n_e}$, and n_i inhibitory inputs represented by $x_{n_e+1}, \ldots, x_{n_e+n_i}$. Any total specification of input is given uniquely by specifying the value of each $x_j = 0$ or 1. There are thus $2^{n_e+n_i}$ possible different total inputs which we can number using a parameter $\alpha = 1, 2, \ldots, 2^{n_e+n_i}$. For example, the total input in which an impulse arrives along every individual input might be that with $\alpha = 1$. It occurs if and only if the truth value of

$$X_1 = x_1.x_2.x_3.\ \ldots\ .x_{n_e+n_i}$$

is 1. For every α there will be a corresponding X_α. For some α the neurone will fire (because $N_e - \phi N_i \geqslant \theta$). Let the set of all such α be called S. Then the expression

$$X = \underset{\text{all } \alpha \text{ in } S}{V} X_\alpha$$

represents the neurone as a truth function.

We now give a worked example in order to clarify the derivation. We consider the following neurone

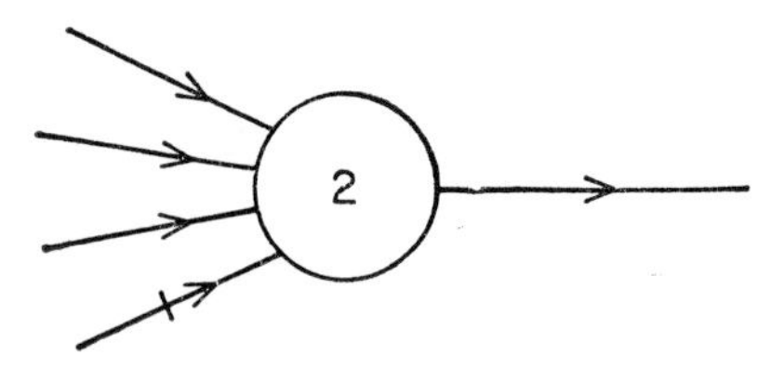

and let $\phi = 1$. Evidently the neurone fires if either two of x_1, x_2 and x_3 fire but not x_4 or if all of x_1, x_2 and x_3 fire, irrespective of whether x_4 does or not. This gives five total inputs which cause the neurone to fire, which we shall number from 1 to 5, with corresponding X_α:

$$X_1 = x_1 . x_2 . x_3 . x_4$$
$$X_2 = x_1 . x_2 . x_3 . \sim x_4$$
$$X_3 = x_1 . x_2 . \sim x_3 . \sim x_4$$
$$X_4 = x_1 . \sim x_2 . x_3 . \sim x_4$$
$$X_5 = \sim x_1 . x_2 . x_3 . \sim x_4$$
$$S = (1, 2, 3, 4, 5)$$
$$X = X_1 V X_2 V X_3 V X_4 V X_5.$$

We can simplify X slightly into the form

$$X = (x_1 . x_2 . x_3) V \{((x_1 . x_2) V (x_1 . x_3) V (x_2 . x_3)) . \sim x_4\}$$

but we see that it is in any case quite complicated even for this neurone which has only 4 inputs.

We have only established the isomorphism for single neurones, but it is easy to extend it by induction to networks without re-entrant paths and, in the other direction, to more complicated logical expressions (which then generally correspond to networks rather than to single neurones). For this and discussion, see McCulloch and Pitts (1943), Kleene (1956).

3.4.2. Switching Circuits

These are used extensively in digital computers and digital equipment generally (for introductory account, see Oppenheimer, 1966). Nowadays they are usually bought as standard modules (complete circuits) each of which gives as its output a definite logical function of its input. Both input and output are standardized electric voltage pulses of height typically a few volts and duration typically from a few nanoseconds to a few micro-seconds. Probably the most extensively used basic units are the AND, OR, NAND and NOR gates. Examples of these follow:

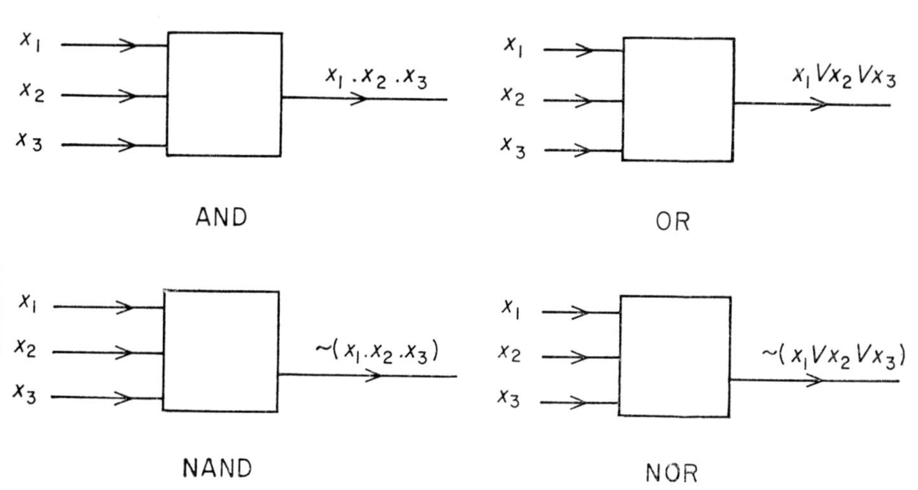

and show the relation between input and output. NAND and NOR gates are more common in practice than one might perhaps expect because of the relative ease of making circuits to perform their functions. It is clear that if we have just two inputs x_1 and x_2 then AND and OR gates perform the basic logical functions $x_1 . x_2$ and $x_1 V x_2$, while with only one input x_1 NAND or NOR perform the function $\sim x_1$. We have seen already that we can build truth functions corresponding to an arbitrary logical neurone and therefore we can also construct a switching circuit to represent it, using these gates, although not necessarily very economically. Incidentally, the AND and OR gates actually shown above represent logical neurones having $n_e = 3$, $n_i = 0$ and $\theta = 3$ and 1 respectively. Finally, note that arbitrary networks of logical neurones, even with re-entrant paths, can be built out of these switching modules. If one was actually doing this, it would be natural to keep the output of all the neurones in the network synchronized using gates controlled by a master clock multivibrator.

3.5. Further Mathematical Features of McCulloch-Pitts Networks

3.5.1. Matrix Formulation of Activity of a Network

We saw in example 5 of Section 3.1.2 that the present state of a network could be conveniently represented as a vector having 0's and 1's for its elements. The threshold condition for a logical neurone involves the expression $N_e - \phi N_i$ which is related linearly to this vector and as a consequence the development of activity in a network can be written in matrix form (this possibility has been pursued particularly by Caianiello and his collaborators, see Caianiello, 1961; Caianiello, de Luca and Ricciardi, 1968; see also Griffith, 1967a, pp. 61–63). We now give a simplified matrix formulation applying to a network of n logical neurones with input from m sensory neurones but without output (this is easily included) and assuming we start at time $t = 0$. Then we have:

1. An n-vector $\mathbf{x}(0)$ with all elements 0 or 1, called the initial vector. For convenience we take this (and all other vectors mentioned) to be a column vector. There are 2^n different possible initial vectors.

2. A fixed n-vector $\boldsymbol{\theta}$, called the threshold vector. All elements θ_i are non-negative integers.

3. A non-linear operator *thr* operating on an n-vector, $\mathbf{v}$ say, and giving a new n-vector *thr* $\mathbf{v}$ according to the rules

$$\begin{aligned} (thr\ \mathbf{v})_i &= thr\ v_i \\ thr\ v_i &= 1 \quad \text{if } v_i \geqslant \theta_i \\ thr\ v_i &= 0 \quad \text{if } v_i < \theta_i \end{aligned}$$

4. A sequence of m-vectors $\mathbf{s}(t)$, $t = 0, 1, 2, \ldots$, called the sense-input vectors. Note that we have taken $t_0 = 0$, $\tau = 1$ (cf. Section 3.1.1).

5. A fixed $n \times n$ matrix C with elements c_{ij} all integers, called the connectivity matrix.

6. A fixed $n \times m$ matrix Q with elements q_{ij} all integers, called the input matrix. $m \geqslant 0$.

7. A relation $\mathbf{x}(t) = thr\,\{Q\mathbf{s}(t-1) + C\mathbf{x}(t-1)\}$, to generate a sequence of n-vectors from the initial vector $\mathbf{x}(0)$. We are interested in the relation between $\mathbf{x}(t)$ and $\mathbf{x}(0)$. If the network has no input, i.e. either $m = 0$ or all $\mathbf{s}(t) = 0$, then

$$\mathbf{x}(t) = (thr\ Q)^t \mathbf{x}(0)$$

but because of the non-linearity of the operator *thr*, this expression is not very tractable.

Remarks. Note that $x_i(t) = 1$ if and only if

$$\sum_{j=1}^{m} q_{ij} s_j(t-1) + \sum_{k=1}^{n} c_{ik} x_k(t-1) \geqslant \theta_i$$

This is just the McCulloch-Pitts kind of threshold relation for neurone i if that neurone has inputs from neurone k as follows:

When $c_{ik} > 0$, there are c_{ik} excitatory inputs;
When $c_{ik} = 0$, there are no inputs;
When $c_{ik} < 0$, there are $-\phi^{-1} c_{ik}$ inhibitory inputs.

The matrix element q_{ij} is similarly interpreted in terms of inputs to neurone i from sensory cell j. We note that ϕ did not appear at all in the matrix formulation and can be chosen to be either $\phi = 1$ or any other value, if such exists, which divides all negative c_{ik} and q_{ij}. It is also clear that if, for example, $c_{ik} \geqslant 0$, we could alternatively have supposed that in the corresponding McCulloch-Pitts network the neurone i has $c_{ik}+a\phi$ excitatory and a inhibitory inputs from neurone k, for any positive integer a, although that would give a less economical realization of the mathematical scheme. For these reasons, the relationship between networks of logical neurones and the mathematical matrix schemes is many: one. However, rules like (1)–(7) above give a self-contained definition of a network of logical neurones and could be taken as a basis for our theory in place of the more heuristic one given in Section 3.1.

Next, we can see the Markovian character (see e.g. Gnedenko, 1962, p. 125) of a McCulloch-Pitts network particularly clearly here, because $\mathbf{x}(t)$ depends directly only on $\mathbf{x}(t-1)$. All memory of earlier states is irrelevant except inasmuch as they have already affected the immediately preceding one.

Finally, as an example, the connectivity matrix for example 5 of Section 3.1.2 is (putting rows and columns in the order A, B, C):

$$C = \begin{bmatrix} 0 & 1 & 1 \\ 1 & 0 & 1 \\ 1 & -1 & 0 \end{bmatrix}.$$

3.5.2. Duality Property of Networks

At any time the state of a network is given by specifying which neurones are firing at that time. Equivalently, however, it would be given if we specify which neurones are not firing. We know that the firing activity develops according to the rules of the neural network. Does the "non-firing" activity develop as if it was in a neural network too? The answer to this is that it does but the new network has different thresholds: we may call it the dual network. We now determine this dual network assuming θ and ϕ are integral. Write $N_e' = n_e - N_e$ and $N_i' = n_i - N_i$ for the numbers of inactive inputs to a neurone, and ask when the neurone does *not* fire in terms of these numbers N_e' and N_i'.

It fires if

$$N_e - \phi N_i \geqslant \theta$$

so it does not fire if

$$N_e - \phi N_i < \theta$$

i.e. if

$$N_e - \phi N_i \leqslant \theta - 1$$

which is if

$$(n_e - N'_e) - \phi(n_i - N'_i) \leqslant \theta - 1$$

or

$$N'_e - \phi N'_i \geqslant n_e - \phi n_i - \theta + 1.$$

So the dual network is obtained by replacing the thresholds θ with $n_e - \phi n_i - \theta + 1$. A self-dual network occurs when these two are equal, i.e. when $\theta = \frac{1}{2}(n_e - \phi n_i + 1)$ for all neurones. There is an analogy here, of course rather superficial, with the hole-particle relation so familiar to physicists in shell models of nuclei, atoms and molecules (see, e.g. Racah, 1942; Griffith, 1964, Section 9.7).

CHAPTER 4

Time Series of Action Potentials

4.1. Microelectrode Recordings

Two kinds of electrical recordings from single cells are common. The first is the intracellular recording, usually made with very fine glass pipettes filled with concentrated KCl solution and having a tip size of 0.5 μm or less (see Katz, 1966, p. 31 for introduction; Donaldson, 1958, Nastuk, 1963, 1964, and Feder, 1968, for further reading on recording techniques). The pipette penetrates the cell wall which, with large neurones, usually closes round it without apparent lasting effect due to the damage, and measures the intracellular potential. However, it is difficult to apply this technique to small cells, which are likely to be badly damaged by the penetration. A much easier procedure is to use a larger microelectrode, usually either a glass pipette or a tungsten or stainless steel wire insulated almost to its end, with a tip diameter of 1 μm or more. Such electrodes are called extracellular because they record the change of local potential outside a cell as a result of the development of an action potential. Naturally the relation between such a measurement and the internal potential, as shown in Fig. 2.3, is rather complicated and depends upon the precise position of the electrode relative to the cell. Nevertheless, in favorable circumstances, the potential changes resulting from the action potential rise far above any background noise and are recognizably reproducible from one firing to the next. See Fig. 4.1 for such a record obtained by G. Horn using a stainless steel microelectrode in the midbrain of an anaesthetized rabbit. Such records are easily and routinely obtainable by an experienced worker by moving the microelectrode through the tissue and stopping when a clear signal appears. However, unlike the intracellular recording, they only show when the cell fires and do not give information about the various EPSP's and IPSP's which underlie the firing.

When an extracellular recording is being made, the experimenter does not normally know exactly which neurone he is recording from but merely the general region in which the cell lies. He may, however, identify the cell functionally by applying various stimuli to the animal and seeing whether the cell fires in response to these. Such experiments have led to a great advance of our understanding of the processing of sensory stimuli, the best-known work on these lines relating to the visual system (see, for example, Lettvin, Maturana, McCulloch and Pitts, 1959; Hubel and Wiesel, 1959, 1965; Barlow, Hill and Levick, 1964; Creutzfeldt, Fuster, Herz and Straschill, 1966).

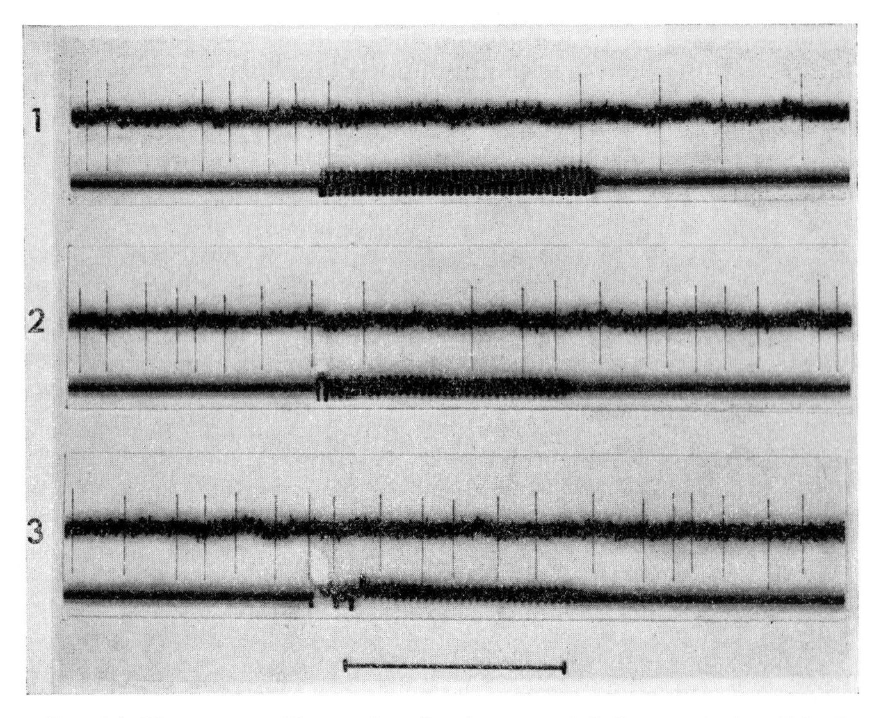

FIG. 4.1. Upper trace: Time series of action potentials from a single cell in the midbrain of an anaesthetized rabbit obtained by extracellular recording with a stainless steel microelectrode, amplified and photographed from an oscilloscope screen. Lower trace: electrical stimulus. Horizontal bar represents 1 second, amplitude of spike potentials about 0.3 mV. The three records represent an uninterrupted sequence. (Recording and photograph by G. Horn, private communication.)

However, even without any such deliberate stimulation, many cells fire continually at generally irregular intervals. Such firing is usually called "spontaneous", although it should not be supposed that there is any necessary implication here that the cell is firing for some entirely intrinsic reason (cf. Section 3.1.2, example 1). There is, no doubt, still input from other cells, but it is not being controlled by the experimenter.

It is relatively easy to obtain very long recordings (e,g. minutes and even hours, with care) of spontaneous activity coming from a single cell (or even occasionally two or more distinguishable cells, see Griffith and Horn, 1963) using a single extracellular electrode. The resulting data consist, then, of a sequence of times, namely the times at which the cell has fired. The question then arises: how is one to interpret these data and can one deduce any important fundamental parameters from them?

4.2. Time Series Analysis

The sequence of times at which a cell fires spontaneously does not normally present any great appearance of regularity. This suggests that we should try to apply statistical methods to its analysis and, using the terminology of statisticians, to regard it as a time series generated by a stochastic process (a process developing in time according to some probabilistic regime). This method has been adopted by several authors (Kuffler, Fitzhugh and Barlow, 1957; Arden and Söderberg, 1961; Werner and Mountcastle, 1963; Gerstein and Mandelbrot, 1964; Levick and Williams, 1964; Griffith and Horn, 1966).

In adopting such an approach we may well ask first what are the simplest such stochastic processes and whether it is reasonable to expect the sequence of firings to follow one of these. The simplest process of all is called the Poisson process and we describe it first (see also Cramér, 1955; Gnedenko, 1962; Cox and Miller, 1965).

A Poisson process occurs when the probability of the cell firing in any interval of time δt is given by $\lambda\delta t + o(\delta t)$, where λ is a constant and $o(\delta t)$ signifies essentially that it may be neglected compared with $\lambda\delta t$ as $\delta t \to 0$. It is a part of the definition that λ is quite uninfluenced by the preceding firings of the cell. In other words the cell is just as likely to fire now if it has fired 100 times in the last second as if it has not fired then at all: the process has no memory.

We now ask how often we expect the cell to fire in a given period. Start at time 0 and let $p_x(t)$ be the probability that the cell fires exactly x times in the interval from 0 to t inclusive. We can find a formula for $p_x(t)$ inductively by first setting $x = 0$. Evidently

$$p_0(t+\delta t) = P_1 P_2,$$

where P_1 = probability the cell does not fire in the time interval $(0, t)$
$= p_0(t)$

and P_2 = probability the cell does not fire in $(t, t+\delta t)$
$= 1-(\lambda\delta t+o(\delta t))$.

Hence

$$p_0(t+\delta t) = p_0(t) - \lambda p_0(t)\delta t + o(\delta t), \tag{1}$$

where $o(\delta t)$ is now a different small quantity. But by Taylor's expansion:

$$p_0(t+\delta t) = p_0(t)+p_0'(t)\delta t+o(\delta t), \tag{2}$$

where $p_0'(t) = dp_0(t)/dt$. Hence, combining equations (1) and (2) we find $p_0'(t) = -\lambda p_0(t)$ which has the solution

$$p_0(t) = e^{-\lambda t}, \tag{3}$$

because obviously $p_0(0) = 1$.

We can now proceed inductively, finding

$$\begin{aligned} p_x(t+\delta t) &= \text{probability of } x \text{ firings in } (0, t) \\ &\qquad\qquad \times \text{ probability of none in } (t, t+\delta t) \\ &\quad + \text{probability of } (x-1) \text{ firings in } (0, t) \\ &\qquad\qquad \times \text{ probability of one in } (t, t+\delta t) \\ &= p_x(t)(1-\lambda\delta t)+p_{x-1}(t)\lambda\delta t+o(\delta t). \end{aligned}$$

But

$$p_x(t+\delta t) = p_x(t)+p_x'(t)\delta t+o(\delta t),$$

so

$$p_x'(t) = -\lambda p_x(t)+\lambda p_{x-1}(t). \tag{4}$$

This is most easily solved by putting

$$p_x(t) = e^{-\lambda t}\, v_x(t), \tag{5}$$

whereupon equation (4) readily simplifies to

$$v_x'(t) = \lambda v_{x-1}(t). \tag{6}$$

But $v_0(t) = 1$ and $v_x(0) = 0$ for $x > 0$. Hence

$$v_x(t) = \lambda \int_0^t v_{x-1}(t)\, dt, \tag{7}$$

from which we obtain successively

$$v_1 = \lambda t,\ v_2 = \tfrac{1}{2}(\lambda t)^2,\ \ldots,\ v_x = (\lambda t)^x/x!,$$

so

$$p_x(t) = \frac{(\lambda t)^x}{x!}\, e^{-\lambda t}. \tag{8}$$

Thus from the Poisson process we derive (cf. equation 2.6) the Poisson distribution for the number of times the cell fires in a time interval of length t.

One of the easiest things to measure is the distribution of intervals between successive firings (spikes). This can be plotted as a histogram, which is called the "interspike interval histogram" or simply the "interval histogram". We now ask what is the theoretical distribution assuming we have a Poisson process. We write it $I(t)$, meaning that $I(t)\,\delta t$ is the probability that an arbitrary chosen interval lies between t and $t+\delta t$. Evidently we have

$$\begin{aligned} I(t)\,\delta t &= \text{probability cell does not fire in } (0, t) \\ &\qquad \times \text{ probability of firing in } (t, t+\delta t) \\ &= \lambda p_0(t)\,\delta t. \end{aligned} \tag{9}$$

Hence our theoretical formula for the interval histogram is

$$I(t) = \lambda\, e^{-\lambda t}. \tag{10}$$

Of course, we must have

$$\int_0^\infty I(t)\,dt = 1, \tag{11}$$

which is easily verified from equation (10). We also find that the mean interval is

$$\bar{t} = \int_0^\infty tI(t)\,dt = \lambda \int_0^\infty t\, e^{-\lambda t}\,dt = \lambda^{-1}, \tag{12}$$

corresponding to the mean firing rate of λ spikes/sec.

One rather straightforward way in which a Poisson process could occur would be if the impulses arriving at the cell were Poisson and the threshold was such that even a single impulse was always sufficient to fire the cell. This possibility is not quite so artificial as it may appear because of the existence of a general theorem in mathematical probability theory (see Khinchin, 1960, and especially Cox, 1962, p. 72, who actually considers a nerve cell as a possible application) which says that, under quite general conditions, the pooling of a large number of non-Poisson stochastic processes leads to a resultant process which approximates to being Poisson. This means that even if the sequence of arrivals of impulses at one presynaptic knob is not close to Poisson, the sequence of arrivals at all those knobs together (which is what is "seen" by the postsynaptic cell) may nevertheless approximate quite well to a Poisson process, bearing in mind that the total number of knobs could be well into the tens of thousands.

It would also seem that, even if the process underlying the sequence of spikes is not closely Poisson, there should always be a Poissonian element for large intervals between spikes. This is because, a long time t after the cell has last fired, it must surely have "forgotten" exactly when it did. Therefore we might expect the probability of firing to settle down to a constant value for very large t. Hence we expect $I(t)$ to have an exponential tail, i.e. to be of exponential form for sufficiently large t. Realizing this possibility and applying suitable statistical tests to their data on unanesthetized cat visual cortex, Griffith and Horn (1966) found 8 out of 31 cells examined, which could be tested, had approximately exponential $I(t)$ over the whole available range of t, while in 25 out of 41 experiments, $I(t)$ had an exponential tail. In those experiments where the tail, for the range of t

measured, was not exponential, there were usually "too many" long intervals (cf. formulae (23) and (35)).

If we have any process for which $I(t)$ is given by equation (10) then, *providing* that different interspike intervals are independent of one another, we can deduce that the process is Poisson. This is easy to see. Suppose that at time t the cell last fired at time $t'(<t)$. Then the probability that it fires in the interval $(t, t+\delta t)$ is given by the conditional probability that the interval lies between $t-t'$ and $t+\delta t-t'$ subject to the condition that it is $\geqslant t-t'$. This is given by

$$I(t-t')\,\delta t \Big/ \int_{t-t'}^{\infty} I(x)\,dx = \lambda\, e^{-\lambda(t-t')}\,\delta t \Big/ \int_{t-t'}^{\infty} \lambda\, e^{-\lambda x}\,dx = \lambda\,\delta t$$

which proves the result.

However, it may well seem that, even if the process is not actually Poisson, different interspike intervals should often be at least approximately independent of one another. This would be expected most strongly when the development of the action potential by the cell wipes out all memory of the EPSP's and IPSP's (including the underlying chemical transmitter, see Section 3.2.2) which have led to it. Deviations from independence are investigated by examining the values of the serial correlation coefficients for the sequence of observed interspike intervals ($T_1, T_2, \ldots, T_n$ say) which are defined as (Kendall, 1948, p. 402)

$$r_p = A_{0p}(A_{00}A_{pp})^{-\frac{1}{2}}, \tag{13}$$

where

$$A_{\alpha\beta} = (n-p)^{-1}\sum_{i=1}^{n-p} T_{i+\alpha}T_{i+\beta} - (n-p)^{-2}\sum_{i=1}^{n-p}\sum_{j=1}^{n-p} T_{i+\alpha}T_{j+\beta}.$$

The expectation of all the r_p, for $p > 0$, is zero when the intervals are independent of each other. It is often found that they deviate significantly from zero in practice. One possible reason for this is the persistence of the chemical transmitter (Section 3.2.2). Another is that any nerve cell forms part of a network, which must almost certainly have many pathways in it which, at least ultimately, finish back on that same cell. This would introduce correlations between present and future activity. Finally, the activity of a given cell may well depend in part on the general level of activity in the brain or in the part of the brain in which it lies. In fact there might be long-term variations in the level of activity, possibly due to changes in attention on the part of the animal, which would be most apparent in the measured mean firing rate per second, but would also tend to make the r_p different from zero. Such variations have been found in cells of the visual cortex of unanaesthetized cat (Griffith and Horn, 1966) and mean that we must be cautious about assuming that our stochastic process is necessarily stationary, i.e. about assuming that the probability regime underlying it is the same at all times.

Nevertheless, providing the mean firing rate is fairly stable, which may be tested first, it is natural to consider as a first approximation the assumption that different intervals are independent, but without also assuming that $I(t)$ is necessarily exponential. The resulting type of stochastic process is very familiar to statisticians and is called a "renewal" process or alternatively a "recurrent" process (see Cox, 1962). In the next section we show that we can use the mathematical neuronal models of Chapter 3 to derive theoretical expressions for $I(t)$.

4.3. Theory of the Interspike Interval Distribution

Throughout this section we shall make the assumption of independence for intervals and shall also suppose that both the arrival of EPSP's and of IPSP's at a cell form Poisson processes (perhaps at two different mean rates). We have seen already that these are not unreasonable approximate assumptions.

4.3.1. Random Walk Model

It is convenient to lead into the discussion by considering the random walk model of Gerstein and Mandelbrot (1964). This is based at first on an assumption of quantization of time, but quite apart from this it has serious defects (Stevens, 1964) which we shall finally remove in Section 4.3.3.

For mathematical simplicity we change the time scale so that the allowed quantized times are $t = \ldots, -2, -1, 0, 1, 2, \ldots$. Then the physical idea is that after the cell fires the internal potential returns to its resting value and, due to the arrival of a random stream of impulses at the cell, has a probability p, at each quantized time, of rising towards the threshold and a probability $q = 1-p$ of going away from it. In the simplest form, we measure the distance from the resting potential as x, which can also take only integral values (e.g. corresponding to the arrival of one EPSP or one IPSP each time). Thus x also $= \ldots, -2, -1, 0, 1, 2, \ldots$. x starts at 0, which is the resting potential, and proceeds by a random walk until it reaches $x = \theta$ when the cell fires. θ is a fixed constant, namely the threshold. Such a walk to $x = m < \theta$ is illustrated in Fig. 4.2a. It is clear that we have merely to calculate the probability that the walk terminates (because $x = \theta$) at a particular value of t to obtain a theoretical formula for the interspike interval distribution.

We now derive this formula (following Gnedenko, 1962). It is easiest to start by taking the case $p = q = \frac{1}{2}$, from which the result for general p can be deduced later. We have what is called a "restricted" random walk problem, the restriction being that the walk stops when we reach $x = \theta$. This value of x is called an "absorbing barrier" in the theory. If the walk were unrestricted (e.g. if $\theta = \infty$), in order to get exactly to $x = m$ at time t (i.e.

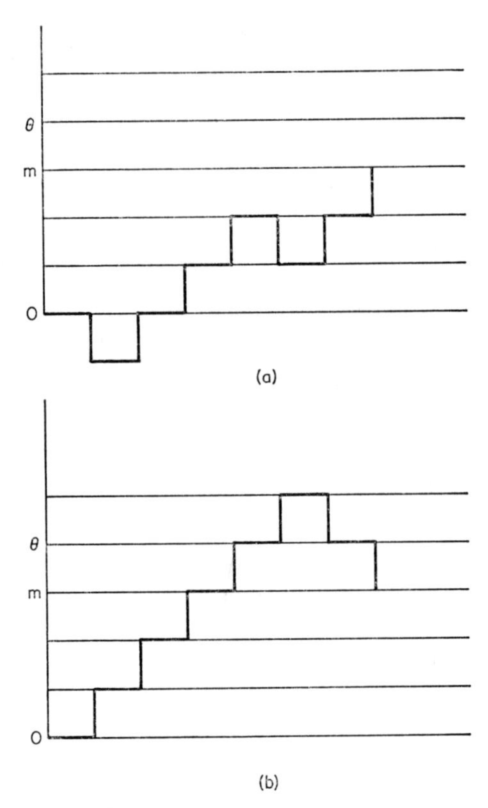

FIG. 4.2. Examples of (a) allowed and (b) not allowed pathways to $x = m$ in the random walk. At each quantized time t, x goes from one value, x_1 say, to $x_1 \pm 1$. We arbitrarily adopt the convention that the value of x actually at t is the second of these two. $x = 0$ for $0 \leqslant t < 1$.

after a total of t steps up and down) we need a path with a positive steps and b negative steps, where a and b must satisfy

$$a+b = t \qquad\qquad a = \tfrac{1}{2}(t+m)$$

whence

$$a-b = m \qquad\qquad b = \tfrac{1}{2}(t-m).$$

There are $\binom{t}{a}$ different ways of doing this when $t+m$ is even and none at all when $t+m$ is odd. Each individual path has probability $(\frac{1}{2})^t$ of occurring, because at each time there is a probability of $\frac{1}{2}$ of taking the appropriate step for the particular path. Hence the probability that $x = m$ at time t is

$$P_t(x=m) = \binom{t}{a}(\tfrac{1}{2})^t = \binom{t}{\frac{1}{2}(t+m)}(\tfrac{1}{2})^t, \quad (t+m \text{ even})$$
$$= 0, \quad (t+m \text{ odd}). \tag{14}$$

We now use a neat trick to deduce the solution of the restricted random walk problem, in which we are interested. We call paths "allowed" if they reach $x = m$ $(m < \theta)$ at time t without having previously reached $x = \theta$ (see Fig. 4.2). Then the trick consists of noticing that from each path to $x = m$ at time t, which is not allowed, we can construct a unique path (which is obviously also not allowed) to $x = 2\theta - m$. This is obtained by reflecting that part of the path which lies between the first point at which $x = \theta$ and the end, in the horizontal line $x = \theta$, and is illustrated in Fig. 4.3

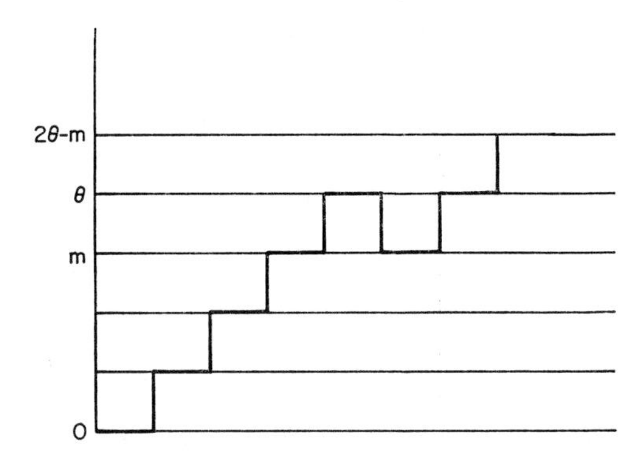

FIG. 4.3. The path obtained by reflecting the latter part of the walk, shown in Fig. 4.2b, in the line $x = \theta$.

for the particular path shown in Fig. 4.2b. Each path from 0 to $x = 2\theta - m$ at time t comes from just one of our original paths which are not allowed. Therefore the total number of the latter paths is the same as the total number of paths to $x = 2\theta - m$ at time t in the unrestricted problem, i.e. it is $P_t(x = 2\theta - m)$, whence it follows that the number of allowed paths is given by

$$\bar{P}_t(x=m) = P_t(x=m) - P_t(x=2\theta-m)$$
$$= \left[\binom{t}{\frac{1}{2}(t+m)} - \binom{t}{\frac{1}{2}(t+2\theta-m)}\right](\tfrac{1}{2})^t, \quad t+m \text{ even}, \tag{15}$$

If the probabilities for the steps up and down are unequal, the number of allowed paths still remains the same, so we have merely to replace $(\frac{1}{2})^t$ in equation (15) with $p^a q^b$ to obtain the formula which is now correct:

$$\bar{P}_t(x=m) = \left[\binom{t}{\frac{1}{2}(t+m)} - \binom{t}{\frac{1}{2}(t+2\theta-m)}\right] p^{\frac{1}{2}(t+m)} q^{\frac{1}{2}(t-m)}, \quad t+m \text{ even}, \tag{16}$$

We now deduce a formula for the interval distribution. Clearly if the cell first reaches threshold at time t, it must have reached $x = \theta-1$ at time $t-1$ by an allowed path and then x must have moved up one unit at time t. Let the probability that this happens be $I_\theta(t)$. Then

$$I_\theta(t) = \bar{P}_{t-1}(x=\theta-1) \times p. \tag{17}$$

The expression in square brackets in equation (16) with $t-1$ for t and $m = \theta-1$ simplifies, with a little elementary algebra, to

$$\frac{\theta}{t}\binom{t}{\frac{1}{2}(t+\theta)}$$

and so

$$I_\theta(t) = \frac{\theta}{t}\binom{t}{\frac{1}{2}(t+\theta)} p^{\frac{1}{2}(t+\theta)} q^{\frac{1}{2}(t-\theta)}, \tag{18}$$

when $t+\theta$ is even and $t \geqslant \theta$, and is zero otherwise.

We next ask whether paths necessarily ever reach the threshold? Clearly it is not logically necessary that a path ever should; it could for example alternate between the value $x = 0$ and $x = -1$ for ever. However, we can still see if the probability of reaching the threshold at any finite future time is unity or less. This means we require to determine the value of the sum

$$S(\theta) = \sum_{t=0}^{\infty} I_\theta(t). \tag{19}$$

We can tackle this by an inductive procedure. First of all, if $\theta = 0$, all paths have length zero because they already start at $x = \theta = 0$. Hence $I_0(0) = 1$, $I_0(t) = 0$ $(t > 0)$ and so $S(0) = 1$. Now consider $S(\theta)$. It is the probability that a path starting at a given value of x (actually $x = 0$) *ever* gets to a time when x has increased by θ. But this probability is equal to the sum of the probability that $x = 1$ at time $t = 1$ and *ever* rises by $\theta-1$, plus the probability that $x = -1$ at time $t = 1$ and *ever* rises by $\theta+1$. In symbols, this reads

$$S(\theta) = pS(\theta-1) + qS(\theta+1). \tag{20}$$

Now, put $\theta = 1$, use $S(0) = 1$ and deduce that

$$S(1) = p + q\,S(2).$$

On the other hand, in a similar way, $S(2)$ is the sum over all t of the probability that a path reaches $x = 1$ at time t and at some future time reaches $x = 2$, i.e.

$$S(2) = \sum_{t=0}^{\infty} I_1(t)S(1) = \{S(1)\}^2.$$

Incidentally, similarly $S(\theta) = \{S(1)\}^\theta$. We now eliminate $S(2)$ and obtain

$$S(1) = p + q\{S(1)\}^2,$$

whence

$$S(1) = 1 \text{ or } p/q. \tag{21}$$

We shall call this $S_p(1)$ to show its dependence on p.

Clearly, if $p = \frac{1}{2}$, $p/q = 1$, so $S_{\frac{1}{2}}(1) = 1$. If $p > \frac{1}{2}$, $p/q > 1$, which is impossible. Hence $S_p(1) = 1$ for all $p \geqslant \frac{1}{2}$. When $p < \frac{1}{2}$, we use equation (18) to obtain

$$\begin{aligned} S_p(1) &= \sum_{t \text{ odd}} \frac{1}{t} \binom{t}{\frac{1}{2}(t+1)} p^{\frac{1}{2}(t+1)} q^{\frac{1}{2}(t-1)} \\ &= \frac{p}{q} \sum_{t \text{ odd}} \frac{1}{t} \binom{t}{\frac{1}{2}(t+1)} q^{\frac{1}{2}(t+1)} p^{\frac{1}{2}(t-1)} \\ &= \frac{p}{q} S_q(1) = \frac{p}{q}, \end{aligned} \tag{22}$$

because $q = 1 - p > \frac{1}{2}$, so $S_q(1) = 1$. We have shown that all paths, with probability unity, reach threshold providing $p \geqslant \frac{1}{2}$. However, if $p < \frac{1}{2}$ only a fraction $S(\theta) = (p/q)^\theta$ ever do so.

We now remark that when the number of steps θ needed to reach threshold becomes very large, formula (18) is approximated by a simpler continuous function. As we must have $t \geqslant \theta$ before we can possibly rise to the threshold, then t also must be large. In these circumstances we know that the binomial expression

$$\binom{t}{\frac{1}{2}(t+\theta)} p^{\frac{1}{2}(t+\theta)} q^{t-\frac{1}{2}(t+\theta)},$$

which is part of formula (18), is well approximated by the normal function (see Cramér, 1955, p. 98):

$$\frac{1}{\sqrt{2\pi t p q}} \exp\{-[\tfrac{1}{2}(t+\theta) - pt]^2/2tpq\}$$

and therefore equation (18) simplifies to

$$I_\theta(t) = \frac{\theta}{\sqrt{8\pi p q}} t^{-\frac{3}{2}} \exp\{-[\theta - (2p-1)t]^2/8pqt\}, \tag{23}$$

where $I_\theta(t)$ is a probability density in the sense that the probability that the interspike interval lies between t and $t+\delta t$ is given approximately by $I_\theta(t)\,\delta t$, providing δt is large compared with unity and small enough that $I_\theta(t)$ does not vary much in $(t, t+\delta t)$. There is a factor of $\frac{1}{2}$ between equations (23) and (18) because the $I_\theta(t)$ of equation (18) is zero whenever $t+\theta$ is odd. With this new expression the probability that the path ultimately reaches threshold is given by integrating $I_\theta(t)$ and is (Bateman Manuscript Project, 1953, vol. 1, 146):

$$P = \int_0^\infty I_\theta(t)\,dt = 1, \qquad p \geqslant \tfrac{1}{2}$$
$$= \exp\,[(p-q)\theta/2pq], \quad p < \tfrac{1}{2} \tag{24}$$

and the mean interval is (same reference):

$$\bar{t} = \int_0^\infty tI_\theta(t)\,dt = \frac{\theta}{2p-1}, \quad p \geqslant \tfrac{1}{2}. \tag{25}$$

The apparent discrepancy between equation (24), with $p < \frac{1}{2}$, and our previous formula for $S(\theta)$ will be resolved when we have seen in the next subsection that the approximation (23) really only works satisfactorily when $p \approx \frac{1}{2}$ and then

$$\exp\,[(p-q)/2pq] \approx \exp\,[2(p-q)] \approx 1+2(p-q) \approx \frac{p}{q}.$$

4.3.2. Connection with Diffusion Equations

When the units of time and of distance from the resting potential used in Section 4.3.1 are small, as measured in seconds or mV, the random walk situation may be described approximately by diffusion equations (see Gnedenko, 1962, p. 329). We see this by introducing new units, h for each step of potential and τ for each step of time, so that the real potential height is $x = x_0 h$ and the real time $t = t_0\tau$. x_0 and t_0 are integral and correspond to the x and t used in the last section.

In an unrestricted random walk we must have

$$P_{t+\tau}(x = x_0 h) = pP_t(x = x_0 h - h) + qP_t(x = x_0 h + h).$$

Let us write $f(x, t)$ in place of $P_t(x)$. Then

$$f(x, t+\tau) = pf(x-h, t) + qf(x+h, t). \tag{26}$$

If we subtract from this the identity

$$f(x, t) = pf(x, t) + qf(x, t)$$

and expand by Taylor's series to the second order we obtain

$$\tau\frac{\partial f}{\partial t} + \tfrac{1}{2}\tau^2\frac{\partial^2 f}{\partial t^2} = p\left(-h\frac{\partial f}{\partial x} + \tfrac{1}{2}h^2\frac{\partial^2 f}{\partial x^2}\right) + q\left(h\frac{\partial f}{\partial x} + \tfrac{1}{2}h^2\frac{\partial^2 f}{\partial x^2}\right)$$

and so

$$\frac{\partial f}{\partial t} + \tfrac{1}{2}\tau\frac{\partial^2 f}{\partial t^2} = -\frac{h(p-q)}{\tau}\frac{\partial f}{\partial x} + \frac{h^2}{2\tau}\frac{\partial^2 f}{\partial x^2}. \tag{27}$$

We are supposing that both h and τ are small and so we now let $h \to 0$, $\tau \to 0$. We only get a sensible result if we also make $p, q \to \frac{1}{2}$, so that both terms on the right of equation (27) can tend to limits. Write $c = \lim\,(h\tau^{-1}(p-q))$

and $D = \lim (h^2/2\tau)$, and we then have

$$\frac{\partial f}{\partial t} = -c\frac{\partial f}{\partial x} + D\frac{\partial^2 f}{\partial x^2}, \tag{28}$$

which is the equation of one-dimensional diffusion with drift (the term $-c(\partial f/\partial x)$).

Gerstein and Mandelbrot (1964) first considered a random walk type of model for the interval distribution and used equation (28) to obtain a theoretical expression for it. It should be pointed out, however, that, although it is natural to simplify the real problem by passing to this continuous limit, equation (28) and the treatment in Section 4.3.1 involve related approximations, and neither is obviously better than the other. In Section 4.3.1 we unjustifiably quantized time and here we unjustifiably suppose the unit PSP's are infinitesimal.

Now return to equation (28). The meaning of $f(x, t)$ is that $f(x, t)\,\delta x$ is the probability at time t that the measure x of the deviation from the resting potential lies between x and $x+\delta x$. Therefore at time $t = 0$ this probability is entirely peaked at the value $x = 0$, and spreads out subsequently. Put mathematically this means that (cf. Dirac, 1947, p. 58):

$$f(x,0) = \delta(x), \tag{29}$$

but those not familiar with Dirac's delta function may simply refer to the discussion following equation (33). The other boundary condition on $f(x, t)$ is that $f(\theta, t) = 0$ for all θ. The reason for this is that all paths terminate immediately when they reach $x = \theta$, so the probability of being actually at $x = \theta$ is zero.

We now solve equation (28) in the range $-\infty < x \leqslant \theta$ subject to these boundary conditions by first transforming it by writing (Chandrasekhar, 1943):

$$\begin{aligned} E(x,t) &= \exp\left[(cx/2D)-(c^2t/4D)\right] \\ f(x,t) &= U(x,t)E(x,t) \end{aligned} \tag{30}$$

Then

$$\frac{\partial f}{\partial t} = E\frac{\partial U}{\partial t} - \frac{c^2}{4D}EU,$$

$$\frac{\partial f}{\partial x} = E\frac{\partial U}{\partial x} + \frac{c}{2D}EU,$$

$$\frac{\partial^2 f}{\partial x^2} = E\frac{\partial^2 U}{\partial x^2} + \frac{c}{D}E\frac{\partial U}{\partial x} + \frac{c^2}{4D^2}EU,$$

whence on substitution in equation (28), followed by division by E, we obtain

$$\frac{\partial U}{\partial t} = \frac{\partial^2 U}{\partial t^2}. \tag{31}$$

This is the equation for simple diffusion without drift, and for heat conduction, in one dimension and its properties are well known (see Carslaw and Jaeger, 1959, p. 50). It has solutions of the kind

$$U(x,t) = t^{-\frac{1}{2}} \exp\left[-(x-a)^2/4Dt\right]. \tag{32}$$

We need a solution for which $f(\theta, t) = 0$ and hence also for which $U(\theta, t) = 0$. This is obtained by the simple combination of two solutions like (32) and is

$$U(x,t) = t^{-\frac{1}{2}} \{\exp\left[-x^2/4Dt\right] - \exp\left[-(x-2\theta)^2/4Dt\right]\}, \tag{33}$$

which evidently has this property. It is also clear that as $t \to 0$ this function, and hence also $f(x, t)$, becomes entirely peaked around $x = 0$ (and $x = 2\theta$, but we are only concerned with $x \leqslant \theta$).

We have finally to satisfy the condition that the probability of being at $x = 0$ at $t = 0$ is unity. This means we require

$$L = \lim_{t\to 0} \int_{-\infty}^{\theta} f(x,t)\, dt = 1.$$

Near $t = 0$, the only significant contributions to the integral involve the first term in equation (33) and values of x near zero. Hence

$$\begin{aligned}
\int_{-\infty}^{\theta} f(x,t)\, dt &\approx \exp(-c^2t/4D)\, t^{-\frac{1}{2}} \int_{-\infty}^{+\infty} \exp\left[(cx/2D)-(x^2/4Dt)\right] dx \\
&\approx \exp(-c^2t/4D)\, t^{-\frac{1}{2}} \int_{-\infty}^{+\infty} \exp(-x^2/4Dt)\, dx \\
&= \exp(-c^2t/4D)\sqrt{4\pi D}
\end{aligned}$$

and so $L = \sqrt{4\pi D}$ and therefore the correctly normalized solution is

$$f(x,t) = \frac{1}{\sqrt{4\pi Dt}} \exp\left[(cx/2D)-(c^2t/4D)\right] \times \{\exp(-x^2/4Dt) - \exp\left[-(x-2\theta)^2/4Dt\right]\}. \tag{34}$$

The flow of paths across $x = \theta$ is given by $\left[-D(\partial f/\partial x)\right]_{x=\theta}$, and this therefore gives our theoretical expression for $I_\theta(t)$. There is no contribution from the term in c because $f = 0$ at $x = \theta$. When $x = \theta$, $U = 0$, so with a little algebra:

$$\begin{aligned}
I_\theta(t) = -D\left(\frac{\partial f}{\partial x}\right)_{x=\theta} &= -\frac{DE}{\sqrt{4\pi D}}\left(\frac{\partial U}{\partial x}\right)_{x=\theta} \\
&= \frac{\theta}{\sqrt{4\pi D}}\, t^{-\frac{3}{2}} \exp\left[-(\theta-ct)^2/4Dt\right].
\end{aligned} \tag{35}$$

This matches with formula (23) and the definitions of c, D providing we remember that θ and t are in different units in the two equations. Corresponding to equations (24) and (25) we have now

$$\int_0^\infty I_\theta(t)\,dt = 1, \qquad c \geqslant 0$$

$$= \exp(c\theta/D), \quad c \leqslant 0 \tag{36}$$

$$\bar{t} = \int_0^\infty t I_\theta(t)\,dt = \theta/c, \qquad c \geqslant 0. \tag{37}$$

When $c < 0$ (or $p < \frac{1}{2}$ in the previous case) the mean interval is infinite simply because the fraction $1-\exp(c\theta/D)$ of paths not included in $I_\theta(t)$ at all are all of infinite length.

One should also comment that the random walk towards the threshold starts after the absolute refractory period of length R (cf. Section 3.2.1). Therefore it is actually the expression $I_\theta(t-R)$ which should be compared with experiment for some $R > 0$. Of course, $I_\theta(t-R) = 0$ for all $t < R$.

Equation (35) depends essentially on the two ratios $\theta/\sqrt{D}$ and $c/\sqrt{D}$. We would like to know $\theta_0 = \theta h^{-1}$, namely the number of steps from the resting potential to the threshold. But $\theta/\sqrt{D} = \theta_0\sqrt{2\tau}$, so we cannot obtain θ_0 itself by fitting our theoretical formula to experiment. Gerstein and Mandelbrot (1964) found quite good agreement between experimental data for three neurones and equation (35) with values for $(\theta^2/D, c^2/D)$ of (100, 0.066), (84, 0.022) and (400, 1.52), respectively, reckoning time units in milliseconds. However, as the model has a very serious defect, these values for the parameters must be viewed with caution. Incidentally one should also mention that our basic assumption throughout this section that the times of arrival of different input impulses are uncorrelated would be invalidated if several knobs were terminal branches of the same axon. Naturally, this will often be the case and therefore θ_0 should probably rather be thought of approximately as the threshold number of input neurones rather than the threshold number of individual knobs.

The most unrealistic feature of the model is its assumption that the PSP can move up or down in response to incoming stimuli, but has no intrinsic tendency to move back again toward the resting potential. This is equivalent to supposing that $\varepsilon = 0$ in equation (3.1) and can only be a reasonable approximation if the time constant ε^{-1} is actually long compared with the mean interval $\bar{t}$. In fact we probably usually have ε^{-1} a few msec (Eccles, 1964, p. 42) and $\bar{t}$ many msec and often even seconds. It is because of this feature that we get the totally unreasonable prediction in the model that when $p < \frac{1}{2}$ or $c < 0$, a fraction of paths never reach threshold, and in fact for them x typically becomes ultimately indefinitely large and negative, which is quite unacceptable physically. Although Gerstein and Mandelbrot deal with this last criticism by introducing a "reflecting" barrier at some negative value of x, this is an entirely artificial device and the problem should be dealt with by taking $\varepsilon \neq 0$, as we do in the next section.

A mathematically interesting case is the borderline one when $p = \frac{1}{2}$ in equation (23) or $c = 0$ in equation (35). Then we have

$$\int_0^\infty I_\theta(t)\, dt = 1, \qquad \bar{t} = \int_0^\infty t I_\theta(t)\, dt = \infty. \tag{38}$$

Gerstein and Mandelbrot (1964) discuss this case, but I cannot feel that it is likely ever to be of much direct relevance to experimental data because here, no matter how large ε^{-1} may be, it can never be long compared with the mean interval! Incidentally, note also that the mean firing rate (which is always $1/\bar{t}$) for such a cell is zero, so a hypothetical brain which was completely filled with such cells would be completely inactive on average, over a long period.

4.3.3. Inclusion of Time Constant for Postsynaptic Potential

We shall now allow the postsynaptic potential to decay, in the absence of further input, toward the resting potential according to equation (3.1) with $\varepsilon > 0$, and thus remove the major defect of the preceding model (Stevens, 1964; Gluss, 1967; Johannesma, 1968). At first we do this in a rather general form which also removes the smaller defect of assuming that the individual component EPSP's and IPSP's are infinitesimal.

We again let $f(x, t)\, \delta x$ be the probability that a path has reached a point between x and $x+\delta x$ at time t. For simplicity we assume at first that each incoming impulse either raises or lowers x by a fixed amount h, not necessarily small, and that excitatory impulses arrive according to a Poisson process of rate λ_e while inhibitory impulses arrive similarly at rate λ_i.

We now derive an equation for $f(x, t)$. In the absence of incoming impulses, x decays according to equation (3.1), i.e. along the curves $x = Ae^{-\varepsilon t}$ as shown in Fig. 4.4. Therefore over a short time interval δt, there are three ways in which a path starting at time t can finish between x and $x+\delta x$ at time $t+\delta t$. It can start in the range given by M in Fig. 4.4 and stay there, i.e. no incoming impulses. Or, it can start in L and be pushed up into the middle cross-hatched region by the arrival of an excitatory impulse. Or, finally, it can start in U and be pushed down by an inhibitory impulse. The probabilities for these three possibilities are, respectively, $1-\lambda_e\, \delta t-\lambda_i\, \delta t$, $\lambda_e\, \delta t$ and $\lambda_i\, \delta t$.

This gives us the equation

$$\begin{aligned} f(x, t+\delta t)\, \delta x &= f(x\, e^{\varepsilon\delta t}, t)\, \delta x\, e^{\varepsilon\delta t}\,(1-\lambda_e\, \delta t-\lambda_i\, \delta t) \\ &\quad + f(x-h, t)\, \delta x\cdot\lambda_e\, \delta t + f(x+h, t)\, \delta x\cdot\lambda_i\, \delta t, \end{aligned} \tag{39}$$

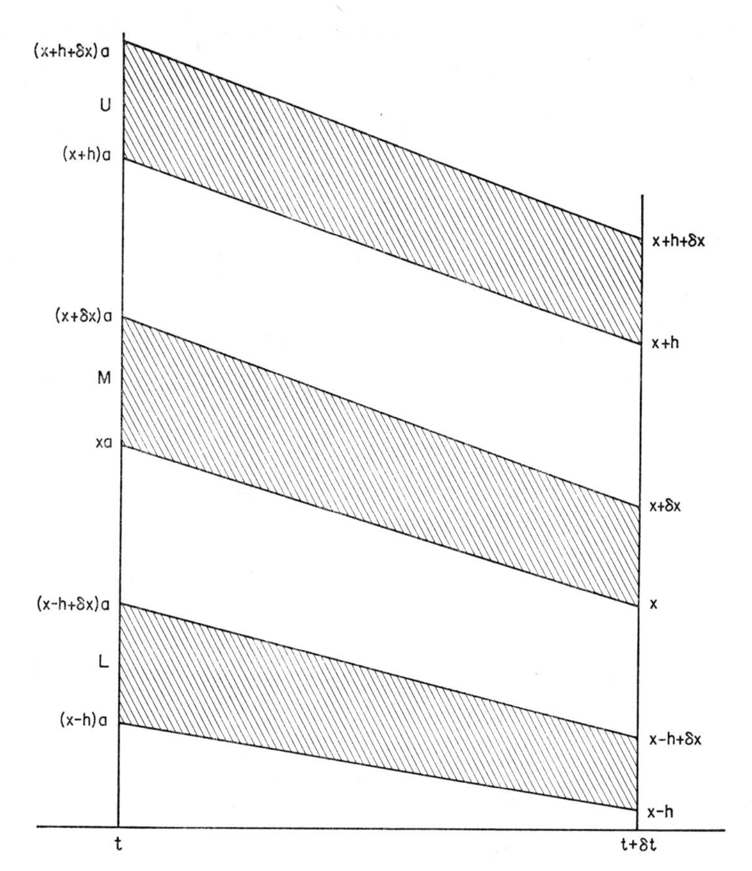

FIG. 4.4. Diagram to illustrate the derivation of the difference equation (39). The curves are $x = Ae^{-\varepsilon t}$ and the parameter $a = e^{\varepsilon \delta t}$.

where we need to retain terms up to $\delta x\, \delta t$. We now expand both

$$f(x\, e^{\varepsilon \delta t}, t) = f(x+\varepsilon x\, \delta t, t) = f(x, t) + \varepsilon x\, \delta t \frac{\partial f(x, t)}{\partial x}$$

and

$$(x, t+\delta t) = f(x, t) + \delta t \frac{\partial f(x, t)}{\partial t}$$

to the appropriate order. On substituting these back into equation (39) and simplifying, we obtain

$$\frac{\partial f(x, t)}{\partial t} = \varepsilon x \frac{\partial f(x, t)}{\partial x} - (\lambda_e + \lambda_i) f(x, t) + \varepsilon f(x, t) + \lambda_e f(x-h, t) + \lambda_i f(x+h, t). \quad (40)$$

This is a differential–difference equation (cf. Section 5.3), which must be solved subject to the boundary conditions

$$f(x, 0) = \delta(x), \quad -\infty < x \leqslant \theta,$$
$$f(x, t) = 0, \quad x \geqslant \theta, \quad \text{all } t.$$

The derivation is essentially unaltered if we let the fixed amount h differ for EPSP's and IPSP's. It just introduces h_e and h_i, respectively, in place of the two h's in equation (40). Indeed, we can generalize further and deal with the case in which there is a continuous distribution of amplitude for the PSP's, to give the equation

$$\frac{\partial f(x, t)}{\partial t} = \varepsilon x \frac{\partial f(x, t)}{\partial x} + \varepsilon f(x, t) - f(x, t) \int_{-\infty}^{+\infty} \lambda(h)\, dh + \int_{-\infty}^{+\infty} \lambda(h) f(x-h, t)\, dh. \quad (41)$$

Naturally $\lambda(h)$ would be zero outside some finite range of values for h. $\lambda(h)$ is fixed, i.e. independent of x and t, and satisfies $\lambda(h) \geqslant 0$ for all h.

Neither equation (40) nor (41) are mathematically very tractable. Consequently we now see what happens when we go back to the approximation of taking h small (equation (40)) or $\lambda(h) = 0$ for h not small (equation (41)). Then we expand

$$f(x \pm h, t) = f(x, t) \pm h \frac{\partial f(x, t)}{\partial x} + \tfrac{1}{2} h^2 \frac{\partial^2 f(x, t)}{\partial x^2},$$

whence we immediately obtain

$$\frac{\partial f}{\partial t} = \varepsilon \frac{\partial}{\partial x}(xf) - c \frac{\partial f}{\partial x} + D \frac{\partial^2 f}{\partial x^2}, \quad (42)$$

where $f \equiv f(x, t)$ everywhere, and

$$c = h(\lambda_e - \lambda_i) \quad \text{or} \quad \int h\lambda(h)\, dh$$
$$D = \tfrac{1}{2} h^2(\lambda_e + \lambda_i) \quad \text{or} \quad \tfrac{1}{2} \int h^2 \lambda(h)\, dh,$$

depending on whether we start from equation (40) or (41).

Gluss (1967) investigated equation (42) and showed that it can be transformed into the simple diffusion equation, as we now see. First we put $f(x, t) = e^{\varepsilon t}\, y(x, t)$, whereupon

$$\frac{\partial y}{\partial t} = (\varepsilon x - c) \frac{\partial y}{\partial x} + D \frac{\partial^2 y}{\partial x^2}$$

and then

$$\xi = \left(x - \frac{c}{\varepsilon}\right) e^{\varepsilon t},$$

$$u = \frac{1}{2\varepsilon}(e^{2\varepsilon t} - 1),$$

whence we obtain

$$\frac{\partial y}{\partial u} = D\frac{\partial^2 y}{\partial \xi^2}.$$

From this, a general solution to equation (42) corresponding to paths starting at $x = w$ at $t = 0$, but with no absorption at $x = \theta$, may be deduced. We use $U(x, t)$ of equation (32), substituting $u = (e^{2\varepsilon t}-1)/2\varepsilon$ for t, $\xi = (x-c/\varepsilon)\, e^{\varepsilon t}$ for x, $w-c/\varepsilon$ for a, remembering that the scale for ξ differs from that for x by a factor of $e^{\varepsilon t}$, to obtain

$$f_w(x, t) = (2\pi D\varepsilon^{-1}(1-e^{-2\varepsilon t}))^{-\frac{1}{2}} \exp\left[\frac{-\{(x-c/\varepsilon)\, e^{\varepsilon t}-(w-c/\varepsilon)\}^2}{2D\varepsilon^{-1}(e^{2\varepsilon t}-1)}\right]. \quad (43)$$

The formula for $f_w(x, t)$ is rather complicated. However, it does enable us to see that there is now no tendency for paths to wander away to infinity. If we let $t \to \infty$ in equation (43) we find that

$$f_w(x, t) \to \left(\frac{\varepsilon}{2\pi D}\right)^{\frac{1}{2}} \exp\left[-\frac{\varepsilon}{2D}\left(x-\frac{c}{\varepsilon}\right)^2\right], \quad (44)$$

which is independent of the starting point, is normalized to unity, and has mean c/ε and standard deviation $(D/\varepsilon)^{\frac{1}{2}}$. It is in fact the steady state solution of equation (42), obtained on putting $\partial f/\partial t = 0$.

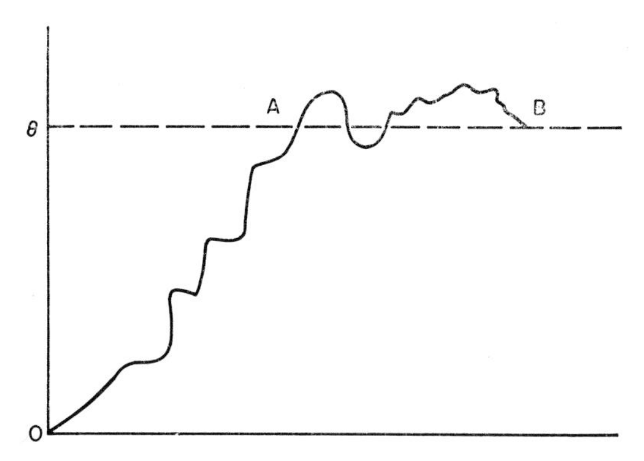

FIG. 4.5. This illustrates the division of an unrestricted path from 0 to B into its first passage from 0 to the line $x = \theta$, followed by the unrestricted path from $x = \theta$ at A back again to $x = \theta$ at B.

Unfortunately it appears that there is no closed solution to equation (42) when we impose the boundary condition $f(\theta, t) = 0$. Therefore $I_\theta(t)$ can presumably be obtained only computationally and compared with experiment. This would be well worth doing, however, as this model may be quite a realistic one, at least for many neurones. We conclude this section by mentioning that although there is no closed simple formula for $I_\theta(t)$, nevertheless

Gluss (1967) has shown that there is an implicit formula in terms of Laplace transforms. This is obtained from the "free" solution of equation (43) by noticing (see Fig. 4.5) that any path in the unrestricted problem going from $x = 0$ at $t = 0$ to $x = \theta$ at time t can be broken up into a first part (0A) which finishes when x first reaches θ and a second part (AB, which could be of zero length) which starts and finishes at θ and is completely *unrestricted* in between. This gives us a relation between the free solutions going from 0 to B and A to B and the restricted solution going to A, as follows:

$$f_0(\theta, t) = \int_{s=0}^{t} I_\theta(s) f_\theta(\theta, t-s)\, ds. \tag{45}$$

Defining the Laplace transforms as

$$L_w(\theta, p) = \int_0^\infty e^{-pt} f_w(x, t)\, dt,$$

$$G(p) = \int_0^\infty e^{-pt} I_\theta(t)\, dt,$$

we then have

$$G(p) = \frac{L_0(\theta, p)}{L_\theta(\theta, p)} \tag{46}$$

which yields $I_\theta(t)$ implicitly. For further discussion of this problem, see Johannesma (1968).

CHAPTER 5

Randomly Connected Networks of Neurones

5.1. Excitatorily Connected Networks of Logical Neurones

5.1.1. Introduction

There is one situation in which we can give a fairly definitive, albeit non-rigorous, theoretical treatment of networks containing arbitrarily many neurones. This is when we have M identical neurones linked together at random, with no input but an arbitrarily assigned initial activity, and has been discussed by various authors (Rashevsky, 1945; Rapoport, 1952; Allanson, 1956; Ashby, Von Foerster and Walker, 1962; Griffith, 1963a). Although it is not to be supposed that any brains are put together at random, it is quite possible that there may exist relatively functionally autonomous parts of brains which consist of one or more pools of neurones each connected more-or-less in this fashion (see especially Burns and Salmoiraghi, 1960, and discussion in Griffith, 1967a, Chapter 5). Furthermore, it is probable that certain general features of neuronal interaction appear particularly clearly here and may be expected to appear in real brains also unless the latter are carefully designed so as to eliminate them.

We shall only treat the case in which all M neurones are identical, i.e. each have the same n_e, n_i, θ and ϕ, but a similar treatment could surely be given if we allowed these parameters to vary. For the purposes of this book, such a generalization would seem to be merely a pointless complication. Furthermore, random linkage could be defined in several different ways but it is not to be expected that the general results which we shall find would alter much if we changed our definition. We shall assume that for each of the n_e+n_i inputs to any neurone, the number $(1, 2, \ldots, M)$ of the neurone from which it comes is chosen independently, with equal probability for each of the M neurones. Thus, in simulating such a network on a digital computer we should have to choose $M(n_e+n_i)$ numbers from the set of integers $1, 2, \ldots, M$, independently and randomly each time. This means that a neurone may be linked to itself or to any other neurone more than once. Also, although each neurone has the same number, i.e. n_e+n_i, of inputs it has a variable number of outputs. This gives a rather simpler theory than the reverse assumption of assigning a fixed number of outputs to each neurone, but it is unlikely that the general behavior of the network would differ very much if we chose the second of these two possibilities (although certain neurones might behave anomalously because they had particularly few input links). Actually, with the first choice, the expected number of outputs per neurone follows the binomial distribution (formula 2.4 with $p = M^{-1}$, $n = M(n_e+n_i)$) and for a large network is well approximated with the Poisson distribution of mean $m = n_e+n_i$ (see formula 2.6).

We must remember however that, although the connections in the network are set up randomly, they remain unchanged thereafter. In other words, on each occasion we are investigating the properties of a quite definite network in which each neurone has inputs from certain other neurones which do not change.

We first consider a network of logical neurones with only excitatory connections and set $n_e = n$, $n_i = 0$ and the parameters $t_0 = 0$, $\tau = 1$ (cf. Section 3.1.1). We now follow Ashby *et al.* (1962) and argue in the following non-rigorous way. At time $t = 0$ some number, x say, of the neurones are active. Write $p = x/M$, which is thus the probability that one of these, selected at random, is active at $t = 0$. Consider any given neurone. Each of its n inputs has the probability p of being active at time $t = 1$. If these n probabilities were independent of one another, we could then say that the probability that exactly N inputs are active at time $t = 1$ is given by the binomial expression

$$\binom{n}{N} p^N(1-p)^{n-N},$$

and hence that the probability that the neurone fires (i.e., $N \geqslant \theta$) at time $t = 1$ is given by

$$P(n, \theta, p) = \sum_{N \geqslant \theta} \binom{n}{N} p^N(1-p)^{n-N}. \tag{1}$$

Our assumption is that this expression is a reasonably accurate approximation to the probability that any given neurone fires at time $t = 1$, and hence also to the fraction of neurones which fire at that time. We thus start with a fraction p active at time $t = 0$ and have a fraction P, according to equation (1), active at $t = 1$. But we can then repeat the procedure and deduce that a fraction $P(n, \theta, P)$ should be active at $t = 2$, and so on. Thus we have an approximate procedure for sequentially calculating the complete future average level of activity of the network.

It is clear that, if we are justified in arguing in this way, the behavior of the network depends crucially on the properties of the function $P(n, \theta, p)$. First, note that if the ratio $R = P/p$ is greater than unity the network is more active at $t = 1$ and if $R < 1$ it is less active. The actual behavior of

P and R as functions of p are shown in Fig. 5.1. It is evident that if $\theta = 1$, $n > 1$, and $p > 0$ at time $t = 0$, then $P > p$ and $P(n, \theta, P) > P$ and so forth. In other words, the activity increases progressively whatever non-zero value of p we start with, and tends towards the fully active state ($P = 1$).

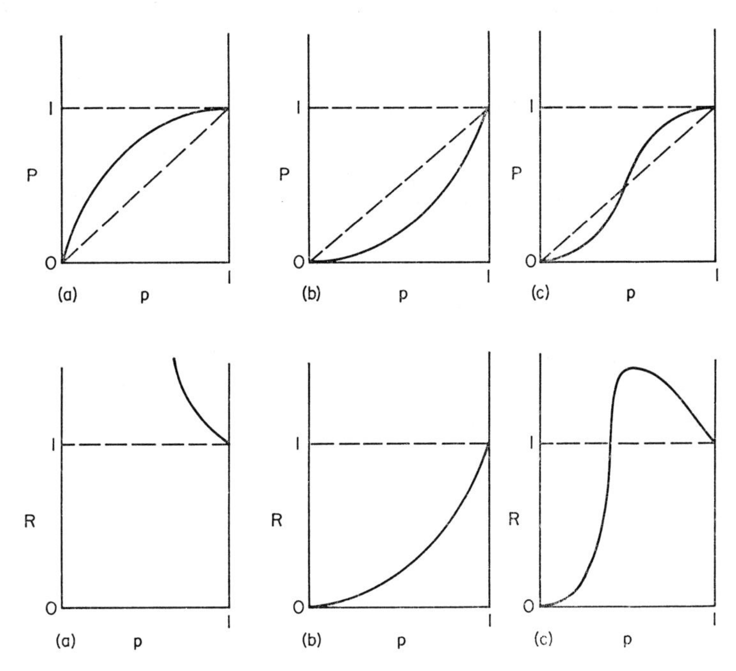

FIG. 5.1. Representative schematic plots (solid lines) of $P(n, \theta, p)$ and $R = P/p$ as functions of p.
(a), $\theta = 1, n > 1$; (b), $\theta = n > 1$; (c), $n > \theta > 1$.

Similarly, if $\theta = n > 1$, $p < 1$, any activity dies away because here $P < p$, $P(n, \theta, P) < P$ and so on. Finally, in the majority of cases $n > \theta > 1$ (the case $\theta > n$ is of no interest because then $P \equiv 0$) and there is an intermediate metastable value for which $0 < p = P < 1$. Any initial activity less than this tends to fade away, while an initial activity greater than it tends to the fully active state. The whole network thus has a threshold to its mean activity, above which the activity is self-maintaining and below which it is not.

5.1.2. PROPERTIES OF $P(n, \theta, p)$

The function $P(n, \theta, p)$ is potentially useful in any situation when we have a McCulloch-Pitts neurone and believe that its total input can be usefully approximated by independent streams of impulses, each of probability p, at each of its input links. This is true even when inhibition is included, as we shall see in Section 5.2. As a consequence it is worth while to investigate the properties of $P(n, \theta, p)$ in detail. In the process we shall prove the assertions we have just made on the basis of the graphs. Certain special cases will be treated first. If $\theta > n$, clearly $P(n, \theta, p) \equiv 0$, while if $\theta = 0$, then

$$P(n, \theta, p) = \sum_{N=0}^{n} \binom{n}{N} p^N (1-p)^{n-N} = (p+(1-p))^n = 1$$

for all n, p. P is relatively simple in three other cases, namely

$$P(1, 1, p) = p$$
$$P(n, 1, p) = 1-(1-p)^n$$
$$P(n, n, p) = p^n$$

which justifies the graphs in Fig. 5.1a and b. None of these cases are likely to have much application, but we give them for completeness. This leaves us with $n > \theta > 1$ and we now establish a number of theorems with application to this range of values for n, θ especially in mind. We assume $n \geqslant \theta \geqslant 1$.

Theorem 1. $P(n', \theta, p) \geqslant P(n, \theta, p)$ whenever $n' \geqslant n$.

Proof. We show this for $n' = n+1$ and can then proceed by induction to $n+2, n+3, \ldots$. We have

$$P(n+1, \theta, p) = \sum_{N=\theta}^{n+1} \binom{n+1}{N} p^N (1-p)^{n-N}.$$

Now

$$\binom{n+1}{N} = \frac{n!}{N!(n+1-N)!}(n-N+1+N) = \binom{n}{N} + \binom{n}{N-1}, \quad N < n+1.$$

Hence

$$P(n+1, \theta, p) = \sum_{N=\theta}^{n} \binom{n}{N} p^N (1-p)^{n+1-N} + \sum_{N=\theta}^{n+1} \binom{n}{N-1} p^N (1-p)^{n+1-N}$$
$$= \sum_{N=\theta}^{n} \binom{n}{N} p^N (1-p)^{n+1-N} + \sum_{N=\theta-1}^{n} \binom{n}{N} p^{N+1} (1-p)^{n-N}$$
$$= \binom{n}{\theta-1} p^\theta (1-p)^{n-\theta+1} + P(n, \theta, p)$$
$$\geqslant P(n, \theta, p).$$

Equality holds only if $p = 0$ or 1 as is evident from the above proof and also from the fact that we always have $P(n, \theta, 0) = 0$, $P(n, \theta, 1) = 1$.

Theorem 2. $P(n, \theta', p) \geqslant P(n, \theta, p)$ whenever $\theta' \leqslant \theta$.

Proof. This is completely trivial because

$$P(n, \theta', p) = \sum_{N=\theta'}^{n} \binom{n}{N} p^N(1-p)^{n-N}$$
$$\geqslant \sum_{N=\theta}^{n} \binom{n}{N} p^N(1-p)^{n-N} = P(n, \theta, p).$$

Theorem 3. $P(n, \theta, p)+P(n, n-\theta+1, 1-p) = 1$.

Proof. This theorem is really true because of the duality discussed in Section 3.5.2. We prove it directly from the definition of P as follows:

$$1 = (1-p+p)^n = \sum_{N=0}^{n} \binom{n}{N} p^N(1-p)^{n-N}$$
$$= \sum_{N=0}^{\theta-1} \binom{n}{N} p^N(1-p)^{n-N} + \sum_{N=\theta}^{n} \binom{n}{N} p^N(1-p)^{n-N}.$$

Then we change the variable of summation in the first sum to $N' = n-N$, and immediately obtain the result.

Corollary. Put $n = 2\theta-1$, then $P(2\theta-1, \theta, p)+P(2\theta-1, \theta, 1-p) = 1$, whence $P(2\theta-1, \theta, \frac{1}{2}) = \frac{1}{2}$, $R(2\theta-1, \theta, \frac{1}{2}) = 1$.

Theorem 4.

$$P(n, \theta, p) = \theta \binom{n}{\theta} \int_0^p x^{\theta-1}(1-x)^{n-\theta}\, dx, 0 < \theta \leqslant n.$$

Proof. We proceed by induction downwards on θ.

$$P(n, n, p) = p^n, \text{ while the R.H.S.} = n \int_0^p x^{n-1}\, dx = p^n,$$

so the theorem is true for $\theta = n$. Now assume it is true for $\theta+1$. Then we integrate by parts and show that it is also true for θ:

$$\theta \binom{n}{\theta} \int_0^p x^{\theta-1}(1-x)^{n-\theta}\, dx$$
$$= \binom{n}{\theta} x^{\theta}(1-x)^{n-\theta}\Big|_0^p + \binom{n}{\theta} \int_0^p x^{\theta}(n-\theta)(1-x)^{n-\theta-1}\, dx$$
$$= \binom{n}{\theta} p^{\theta}(1-p)^{n-\theta}+(\theta+1) \binom{n}{\theta+1} \int_0^p x^{\theta}(1-x)^{n-\theta-1}\, dx$$
$$= \binom{n}{\theta} p^{\theta}(1-p)^{n-\theta}+P(n, \theta+1, p)$$
$$= P(n, \theta, p)$$

which establishes the theorem by induction.

Corollary 1.

$$P' = \frac{dP}{dp} = \theta \binom{n}{\theta} p^{\theta-1}(1-p)^{n-\theta}.$$

Hence $P' \geqslant 0$, with equality only when $p = 0$ or 1.

Corollary 2.

$$P'' = \frac{d^2P}{dp^2} = \theta \binom{n}{\theta} p^{\theta-2}(1-p)^{n-\theta-1}((\theta-1)(1-p)-(n-\theta)p).$$

Hence $P'' = 0$ when $p = 0$ $(\theta > 2)$ or $= 1p$ $(\theta < n-1)$ or $p = (\theta-1)/(n-1)$.

Theorem 5. $P(n, \theta, p) = p$ is true for $p = 0$ or 1. If $\theta = 1$, $n > 1$ or $\theta = n > 1$ it is true for no other value of p. If $\theta = n = 1$, it is true for all p. If $n > \theta > 1$, it is true for $p = 0, 1$ and just one other value p_0 say. $0 < p_0 < 1$.

Proof. All cases except $n > \theta > 1$ follow at once from the formulae for P already given at the beginning of this subsection. The last case can be treated by considering the zeros of the continuous function $Q(p) = P - p$. Clearly $Q(0) = Q(1) = 0$ and from Theorem 4, Corollary 1, $Q'(0) = Q'(p) = -1$. Hence Q has at least one other zero and unless $Q'(p) = 0$ for more than two values of p between 0 and 1, not more than one (see Fig. 5.2a). But from Corollary 2, $P' \geqslant 0$ at $p = 0$,

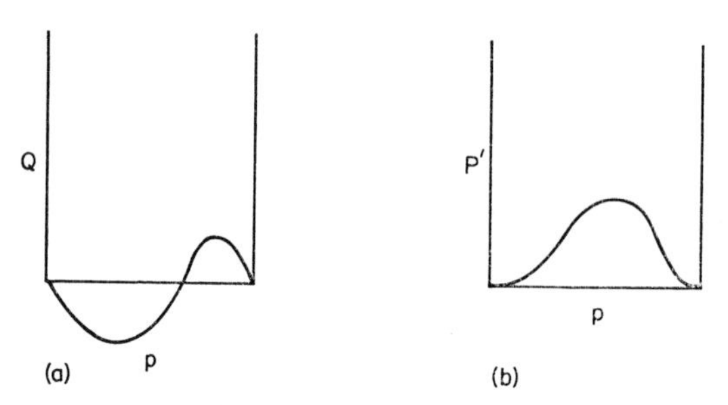

FIG. 5.2. Schematic plots of (a) $Q(p)$ and (b) $P'(p)$ as functions of p.

$P' \leqslant 0$ at $p = 1$ and $P' = 0$ at just the one value in between (see Fig. 5.2b). Hence $Q' = P' - 1$ cannot be zero at more than two values of p. Hence there is just one point p_0, $0 < p_0 < 1$, at which $Q(p_0) = 0$, i.e., at which $P(n, \theta, p_0) = p_0$.

Comment. When $n = 2\theta - 1$, it follows from Theorem 3, Corollary, that

$p_0 = \frac{1}{2}$. More generally, if we write $p_0 = p_0(n, \theta)$ then Theorem 1 shows $p_0(n', \theta) > p_0(n, \theta)$ when $n' < n$, Theorem 2 shows $p_0(n, \theta') > p_0(n, \theta)$ when $\theta' > \theta$ and Theorem 3 that $p_0(n, \theta) + p_0(n, n-\theta+1) = 1$.

Theorem 6.

$$P(n, \theta, p) \approx \frac{1}{\sqrt{2\pi}} \int_a^\infty \exp\left(-\tfrac{1}{2}x^2\right) dx,$$

where

$$a = \left(\frac{\theta}{n} - p\right)\left(\frac{n}{p(1-p)}\right)^{\frac{1}{2}}.$$

This follows at once from de Moivre's theorem (Cramér, 1955, pp. 96–102), and is a good approximation when n is large and θ/n not too small. It means that in those circumstances $p_0 \approx \theta/n$, and that P approximates to the step function

$$P(n, \theta, p) \approx S(\theta n^{-1}, p) = 0, \quad 0 \leqslant p < \theta/n$$
$$P(n, \theta, p) \approx S(\theta n^{-1}, p) = 1, \quad \theta/n \leqslant p \leqslant 1. \tag{2}$$

However, if we keep θ fixed and let $n \to \infty$, we find that p_0 is systematically less than θ/n and we can easily show:

Theorem 7. With θ constant and $n \to \infty$, p_0 is asymptotically equal to

$$(\theta!/n^\theta)^{1/\theta-1}, \quad \theta > 1.$$

Proof. We require to obtain the most significant contribution to the solution of the equation

$$p_0 = P(n, \theta, p_0) = \sum_{N=\theta}^{n} \binom{n}{N} p_0^N (1-p_0)^{n-N}.$$

As this equation can be written as

$$1 = \binom{n}{\theta} p_0^{\theta-1}(1-p_0)^{n-\theta} + \ldots$$

it is clear that p_0 must be small and that the first approximation, for n large, is likely to be given by the first term. Also

$$\binom{n}{\theta} \approx n^\theta/\theta!, \qquad (1-p_0)^n \approx 1,$$

hence $n^\theta p_0^{\theta-1} \approx \theta!$ which gives the result required. It is easy to check, using this p_0, that the neglected terms are in fact of lower order.

Corollary. When $\theta = 2$, $p_0 \sim 2/n^2$. Actually, if $n = 100$, $p_0 = 2.05/10^4$.

Comment. More accurate asymptotic expressions for p_0 when θ is fixed, and a table of some values of p_0 for low θ and n are to be found in Griffith (1967a), pp. 55–58.

We have seen in Theorem 4 that P' has a very simple expression. It is also true that the integral from 0 to 1 of P is simple. In fact, so are all the moments (Cramér, 1955, p. 74) of the distribution as we now show.

Theorem 8.

$$\int_0^1 P(n, \theta, p)p^m \, dp = \frac{1}{m+1}\left[1 - \frac{n!(m+\theta)!}{(\theta-1)!(n+m+1)!}\right].$$

Proof. Write I_m for the integral we wish to evaluate. Then use Theorem 4 and change the order of integration thus:

$$I_m = \theta \binom{n}{\theta} \int_0^1 dp \int_0^p x^{\theta-1}(1-x)^{n-\theta} p^m \, dx$$

$$= \theta \binom{n}{\theta} \int_0^1 dx \int_x^1 x^{\theta-1}(1-x)^{n-\theta} p^m \, dp$$

$$= \frac{\theta}{m+1} \binom{n}{\theta} \left[\int_0^1 x^{\theta-1}(1-x)^{n-\theta} \, dx - \int_0^1 x^{m+\theta}(1-x)^{n-\theta} \, dx\right].$$

The first term is simply $(m+1)^{-1}P(n, \theta, 1) = (m+1)^{-1}$ while the second is obtained from

$$1 = P(n+m+1, m+\theta+1, 1) = \frac{(n+m+1)!}{(m+\theta)!(n-\theta)!} \int_0^1 x^{m+\theta}(1-x)^{n-\theta} \, dx$$

which also uses Theorem 4. After a little rearrangement, the result follows.

Corollary.

$$I_0 = \int_0^1 P(n, \theta, p) \, dp = 1 - \frac{\theta}{n+1}.$$

Compare this with equation (2).

5.1.3. Variability of Network Behavior

At best, the quantity $MP(n, \theta, p)$ gives the expected number of neurones active at time t, if Mp were active at time $t-1$. There will also be a variability in this number depending on the precise details of connection of the particular network concerned and upon exactly which Mp neurones were active at time $t-1$. One approach to this is to remark that at time t, each of the M neurones has a probability P of firing. Hence, the number which actually do so should follow the binomial distribution with mean MP and standard deviation $\sqrt{MP(1-P)}$. Thus the fraction which fires at time t should have a distribution with mean P and standard deviation $\sqrt{P(1-P)/M}$ and thus be sharply peaked for a large network.

We have run a number of computer simulations of McCulloch–Pitts networks of the type considered here and have generally found good agreement with the approximate theory of the expected number of firings. One also notices some variability, although this was not investigated systematically. See Fig. 5.3 for a few illustrative runs of networks having the theoretical quantity $p_0 = \frac{1}{2}$, and $Mp_0 = 50$.

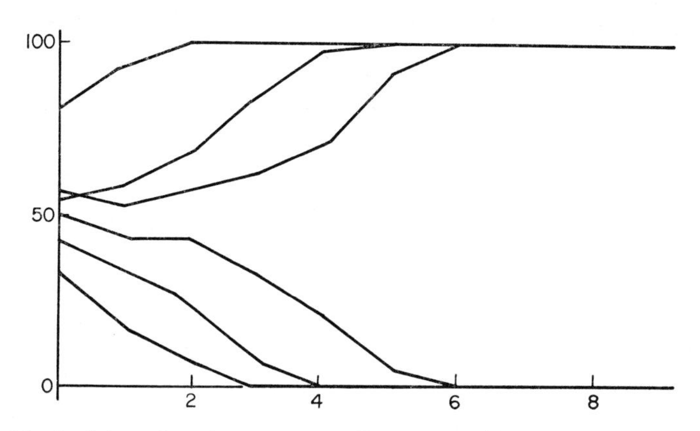

FIG. 5.3. Activity of various masses of 100 McCulloch–Pitts neurones plotted as a function of time. Typical runs starting with various initial activities. $\theta = 3$, $n = 5$, $p_0 = \frac{1}{2}$, $M = 100$.

Of course, we must realize that although the general feature that we have found, namely that the whole network has a threshold of average activity below which it rapidly becomes quiescent and above which it becomes permanently maximally active, usually occurs, it is certainly also possible to construct networks in which it does not. For example, if the network happens to consist of two totally disconnected parts, one part could persist in the inactive state while the other was permanently active. The best one can hope for is that such networks are unlikely to occur when the connections are assigned at random in the manner we have considered. It is possible that some theorem might be proved to the effect that the typical behavior occurs provided that the network is sufficiently "tightly" connected, in some sense. There may be some useful analogy here with the concept of "metric indecomposability" discussed by Khinchin (1949) in connection with the ergodic problem of classical statistical mechanics.

5.1.4. Field-Theoretic Approaches

In thinking about the large array of neurones which forms any natural brain, we must be always on the look-out for approximate mathematical methods of predicting the overall behavior of activity in the array. One natural way to try to proceed is to replace the true, discrete, array with an approximating continuous one and then to try to describe the activity as a field quantity satisfying field equations. In a pioneering work, Beurle (1956) treated activity as a field quantity and, in the same spirit Griffith (1963b, 1965a) showed that fairly simple field equations could be written for a similar quantity, giving what may be called a neural field theory. Both authors were essentially concerned with excitatorily connected networks and the results which they obtained were in general qualitative agreement with those that we have just obtained using the function $P(n, \theta, p)$. We now give a brief outline of the neural field theory, referring to the original papers for a more detailed discussion.

In a field theory we normally have two kinds of elements. The first are the field quantities, satisfying partial differential equations, usually of the second order. The second are the sources (or sinks) of the field, with rules for their functional relationship to the field. The sources are taken to be the bodies of the neurones and their strength at a given time to be proportional to some parameter, probably the mean firing rate at the time, giving a measure of the level of activity of the neurone. We shall denote the strength of the sources as F, which is a function of position and time. The field quantity is the density of action potentials passing from neurones along axons and ultimately, at synapses, having a potentially excitatory effect on other neurones. This quantity we call ψ which again is a function of x, y, z and t. We shall suppose F and ψ to be continuous functions of their arguments and to possess all derivatives up to the second order at least. These continuous functions can, of course, only represent mathematical approximations to the true discrete situation but this is not in itself necessarily objectionable and is in line with the neglect, in continuum mechanics, of atomic structure.

We now consider the general features of the physical quantities correlated with ψ and F and how they are to be preserved in our model. Before doing so, it is well to emphasize that, in the model, excitation is regarded as being carried by a continual shuttling between sources and field. F creates ψ, which in turn influences F at a different point. This new F creates more ψ and so on.

Thus F at a given point P creates ψ which then travels outward, decaying in relation to the number of other neurones with which a typical neurone near P is directly connected. Presuming that impulses travel independently of one another, we can deduce a principle of superposition for our field. Put mathematically, if (F_1, ψ_1) and (F_2, ψ_2) are two solutions of the field equations, it will be natural to demand that $(\lambda F_1+\mu F_2, \lambda\psi_1+\mu\psi_2)$ will at least be solutions of the equations for ψ as functions of F, when λ and μ

are any real constants. This makes us adopt a linear inhomogeneous differential equation for ψ of the form

$$H\psi = F, \tag{3}$$

where H is a linear operator. The independent variables are x, y, z and t. The strength of the sources at a given time depends directly on the local firing activity of the neurones which, in its turn depends on the local level of arrival of impulses, i.e. on ψ at that point. We write

$$F = -4\pi f(\psi),$$

where the factor -4π must seem merely an arbitrary choice here, but arises naturally in the more complete treatment. $f(\psi)$ is generally non-linear (see Fig. 5.4).

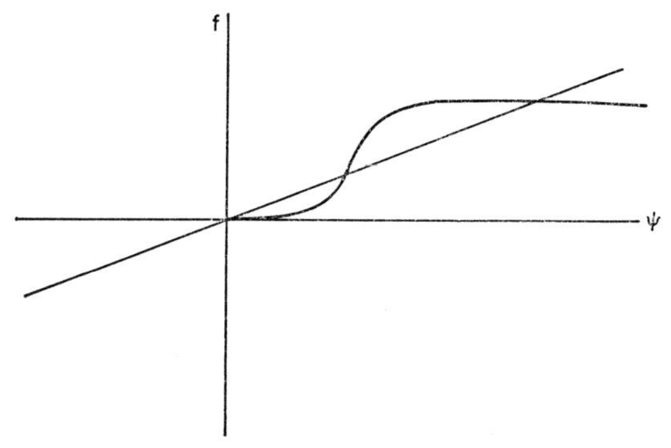

FIG. 5.4. Schematic plot of a possible field-theoretic source distribution $f(\psi)$. The oblique straight line has slope $\beta^2/16\pi$.

We now have our basic equation

$$H\psi = -4\pi f(\psi) \tag{4}$$

and if we wish to apply it to a region which is uniform and has no preferred spatial direction we must take H to be invariant under all translations and rotations.

Assuming H is also a second-order operator this means we must have

$$H = a + b\frac{\partial}{\partial t} + c\frac{\partial^2}{\partial t^2} + d\nabla^2. \tag{5}$$

Consider now the mathematically simple situation of taking one single neurone (or a small group of neurones) at the origin of coordinates, with no others anywhere. Assume spherical symmetry for this neurone (or for the group) with a resultant spherical symmetry for the field ψ. Furthermore, at first, let the firing rate of this neurone, or of these neurones, be constant so that we have

$$\frac{\partial \psi}{\partial t} = \frac{\partial^2 \psi}{\partial t^2} = 0,$$

$$\psi = \psi(r),$$

$$\frac{\partial \psi}{\partial x} = \frac{\partial r}{\partial x}\frac{d\psi}{dr} = \frac{x}{r}\frac{d\psi}{dr},$$

$$\frac{\partial^2 \psi}{\partial x^2} = \frac{1}{r}\frac{d\psi}{dr} - \frac{x^2}{r^3}\frac{d\psi}{dr} + \frac{x^2}{r^2}\frac{d^2\psi}{dr^2},$$

$$\nabla^2 \psi = \frac{2}{r}\frac{d\psi}{dr} + \frac{d^2\psi}{dr^2},$$

and $f(\psi) = 0$ away from the origin because there are no neurones except at $r = 0$. Then equation (4) becomes

$$a\psi + \frac{2}{r}\frac{d\psi}{dr} + \frac{d^2\psi}{dr^2} = 0, \quad r \neq 0,$$

whence

$$ar\psi + \frac{d^2}{dr^2}(r\psi) = 0,$$

$$r\psi = A\, e^{-\frac{1}{2}\beta r},\ \beta > 0,$$

where $a = -\frac{1}{4}\beta^2$ and we have rejected the physically unacceptable term in $e^{+\frac{1}{2}\beta r}$. The interpretation of this is that the field theory is only consistent if we assume that the connections from each neurone directly to other neurones falls off as $r^{-1}\,e^{-\frac{1}{2}\beta r}$. Such a distribution of connections is in fact not too far from what has been observed by Sholl (1956) for some cells in the visual cortex of cat.

We can see how time should be included by taking a variable activity for the neurones at $r = 0$. Then if the velocity of conduction is v, the value of ψ at r at time t should reflect the neural activity at the origin at time $t-r/v$. So with the above connectivity function we should have

$$\psi = A\left(t - \frac{r}{v}\right) r^{-1}\, e^{-\frac{1}{2}\beta r}$$

as a solution of equation (4), where $A(t)$ measures the source activity at the origin as a function of time. As

$$\nabla^2 \psi = \frac{1}{r}\frac{\partial^2}{\partial r^2}(r\psi) = \frac{1}{r}\frac{\partial^2}{\partial r^2}(A e^{-\frac{1}{2}\beta r})$$

$$= \frac{1}{r}\, e^{-\frac{1}{2}\beta r}\left(\frac{\partial^2 A}{\partial r^2} - \beta\frac{\partial A}{\partial r} + \frac{1}{4}\beta^2 A\right)$$

$$= \frac{1}{r} e^{-\frac{1}{2}\beta r} \left(\frac{1}{v^2} \frac{\partial^2 A}{\partial t^2} + \frac{\beta}{v} \frac{\partial A}{\partial t} + \tfrac{1}{4}\beta^2 A \right)$$

$$= \left(\frac{1}{v} \frac{\partial}{\partial t} + \tfrac{1}{2}\beta \right)^2 \psi,$$

this determines a, b and c in terms of d. In fact the constant d can be taken to be unity (Griffith, 1963b, p. 117) and so equation (4) becomes

$$H\psi \equiv -\left(\frac{1}{v} \frac{\partial}{\partial t} + \tfrac{1}{2}\beta \right)^2 \psi + \nabla^2 \psi = -4\pi f(\psi). \tag{6}$$

If, following our discussion of p_0 before, we search for stationary solutions of (6), uniformly distributed in space, we can set

$$\frac{\partial \psi}{\partial x} = \frac{\partial \psi}{\partial y} = \frac{\partial \psi}{\partial z} = \frac{\partial \psi}{\partial t} = 0$$

and see that the constant value $\psi = c$, say, must satisfy the equation

$$\tfrac{1}{4}\beta^2 c = 4\pi f(c).$$

A reasonable form for the function f is shown in Fig. 5.4, which has three stationary points. The function must flatten off for large c because there must be some upper limit to the firing rate of the cells. Figure 5.4 is very suggestive of two stable activities $\psi = c_0, c_2$, with an intermediate unstable one $\psi = c_1$, as we found previously in connection with $P(n, \theta, p)$. There is some evidence that all other initial activities do in fact settle down either to $\psi = c_0$ or $\psi = c_2$, but no rigorous proof of this has been given (Griffith, 1965a).

5.1.5. Real Time Neurones

So far we have only derived the expected behavior of a mass of excitatorily and randomly connected neurones using relatively crude mathematical models for the network. However, there is every reason to suppose that similar properties would emerge for more realistic neuronal models and even for such aggregates of actual nerve cells. We shall now discuss this for real time neurones, using that limiting version of the model which relates the parameter V to the residual transmitter level (cf. Section 3.2.2). Thus V is not reset to zero a period R after a cell has fired. We shall take all increments η (cf. Section 3.2.1) equal and, without loss of generality, put $\eta = 1$.

We set up the connections in the random network as before and suppose that the neurones are firing at a rate of λ/second. The intention is to ask whether this rate can be self-maintaining, and hence whether such an activity would die away or perhaps increase to some maximum rate. First we give an approximate treatment which gives a physical insight into the matter.

Consider the mean value of V for one cell, i.e.

$$\bar{V} = T^{-1} \int_0^T V(t)\, dt \tag{7}$$

over a long time t. Each impulse arriving at one of the n inputs to the cell during this time ultimately contributes to the integral an amount

$$\int_0^\infty e^{-\varepsilon t}\, dt = \varepsilon^{-1}$$

if we let T become large enough (cf. equation 3.1). But on average $n\lambda$ impulses arrive per second and so for large T we find

$$\bar{V} \approx T^{-1} \cdot n\lambda T\varepsilon^{-1} = n\lambda/\varepsilon.$$

Thus $\bar{V}$ should fluctuate around the value $n\lambda/\varepsilon$. So if $n\lambda/\varepsilon$ is much larger than θ the cell will fire as rapidly as possible. Because of the refractory period, this means firing regularly at intervals of R, i.e. at a rate $\lambda = R^{-1}$/second. If $n\lambda/\varepsilon$ is much smaller than θ, the cell will virtually not fire at all, i.e. the activity will immediately die away to zero. Thus the threshold of activity for the mass of cells must lie somewhere near the point at which $n\lambda/\varepsilon = \theta$, i.e. $\lambda = \varepsilon\theta/n$. It is reassuring to find how closely this formula compares with the McCulloch–Pitts network threshold $p_0 \approx \theta/n$.

The state of maximum activity is when each cell fires at the rate $\lambda = R^{-1}$/second. We now examine this state in more detail. The input to any chosen neurone along one of its n inputs consists of a regularly spaced sequence of impulses, R apart. Therefore its contribution to V is given by

$$C = \sum e^{-\varepsilon t_j},$$

where t_j are the times from the arrivals of the equally spaced impulses. Hence ultimately

$$C \to \sum_{j=0}^{\infty} e^{-\varepsilon(jR+\gamma)} = e^{-\varepsilon\gamma}(1-e^{-\varepsilon R})^{-1}, \tag{8}$$

where γ is the time from the arrival of the last impulse along that input link. But $0 \leqslant \gamma < R$ and so we can deduce that

$$e^{-\varepsilon R}(1-e^{-\varepsilon R})^{-1} < C \leqslant (1-e^{-\varepsilon R})^{-1}$$

and therefore that at all times V itself satisfies

$$n(e^{\varepsilon R}-1)^{-1} < V \leqslant n(1-e^{-\varepsilon R})^{-1}. \tag{9}$$

As a cell can only fire if $V \geqslant \theta$, it follows that we have shown that the state of maximum activity (and of course any other state) cannot maintain itself if

$$n(1-e^{-\varepsilon R})^{-1} < \theta. \tag{10}$$

In such a network, any initial activity must fade away to zero. This corresponds to the McCulloch–Pitts case in which $\theta > n$. On the other hand, if

$$n(e^{\varepsilon R}-1)^{-1} \geqslant \theta \tag{11}$$

the state of maximum activity maintains itself indefinitely, no matter how the activities of the constituent neurones may happen to be correlated. Finally, if θ lies between these two values, the outcome is uncertain and the state of maximum activity might usually persist for a while and then ultimately fade away to zero. Note that our previous approximate calculation would have given us a single critical value of $\theta = n/\varepsilon R$ (taking $\lambda = R^{-1}$). This is quite consistent with the present more accurate discussion because it follows from the properties of the exponential function that

$$n(1-e^{-\varepsilon R})^{-1} > n/\varepsilon R > n(e^{\varepsilon R}-1)^{-1}$$

for all $\varepsilon R > 0$.

Naturally if such networks of neurones exist in parts of animal brains, there would have to be methods of switching them off again after they got into the state of maximum activity. One possible way in which this could occur would be through inhibitory impulses arriving along axons of cells lying outside the random network. However, another way is also possible, namely through the habituation of the cells in the network due perhaps to a progressive rise in the threshold θ. We now consider the theory of this, using the model for habituation which was given at the end of Section 3.2.2.

If the mass of cells fires at its maximum rate $\lambda = R^{-1}$, we can calculate the limiting value towards which θ will tend. It is

$$\begin{aligned}\theta_{\text{lim}} &= \theta_0 + h\sum_{j=1}^{\infty} e^{-\beta(jR+\gamma)} \\ &= \theta_0 + h\,e^{-\beta\gamma}(1-e^{-\beta R})^{-1}\end{aligned} \tag{12}$$

in the same way as we derived equation (8). However, here the value of θ_{lim} is only relevant when the cell has just reached the end of its refractory period and is ready to fire again. So we must set $\gamma = R$ in equation (12) to obtain the unique value

$$\theta_{\text{lim}} = \theta_0 + h(e^{\beta R}-1)^{-1}. \tag{13}$$

Comparing this with equations (10) and (11) we can deduce that habituation will ultimately cut the activity off if this θ_{lim} satisfies

$$\theta_{\text{lim}} > n(1-e^{-\varepsilon R})^{-1} \tag{14}$$

but cannot if (equation (15) is incorrectly given in Griffith, 1967a, p. 61)

$$\theta_{\text{lim}} < n(e^{\varepsilon R}-1)^{-1}. \tag{15}$$

As before, there is a range of uncertainty.

A possible situation to which the present theory might be applicable is to the mammalian respiratory system. Here it is believed that there are two sets of neurones, each of which is likely to be excitatorily interconnected within itself, possibly fairly randomly. When one set is active, the animal breathes in, and when the other is, it breathes out. There are inhibitory interconnections to ensure that the two are not simultaneously active. Salmoiraghi and Von Baumgarten (1961) believe from their intracellular recordings that the activity of one set is cut off by a gradually increasing threshold θ of the individual cells as we have discussed here. However, in this case it is unlikely that the neurones behave according to the simpler limiting version of the real time neuronal model which we have used, because they do not fire regularly during their active periods, and this would therefore introduce an extra complexity into the theory.

5.2. Networks with Inhibitory Connections

5.2.1. Extension of Theory

When we allow some of the connections in the random network to be inhibitory, the range of possible behavior for the mean firing activity becomes greater. We shall discuss this for McCulloch–Pitts networks, generalizing our previous treatment.

Let a neurone now have n_e excitatory inputs and n_i inhibitory inputs and let the constants n_e, n_i, θ and ϕ (cf. Section 3.1.1) be the same for each neurone of the network. Again we choose each connection independently and entirely at random from the M neurones. If a fraction p are active at time $t = 0$, it follows in the same way as we discussed in Section 5.1.1 that the expected fraction at time $t = 1$ is given by

$$P_1(p) = \sum_{N_e - \phi N_i \geq \theta} \binom{n_e}{N_e} \binom{n_i}{N_i} p^{N_e + N_i}(1-p)^{n_e + n_i - N_e - N_i}, \tag{16}$$

where the sum is over all pairs of integers N_e, N_i in the ranges $0 \leqslant N_e \leqslant n_e$, $0 \leqslant N_i \leqslant n_i$ which satisfy the inequality shown.

This formula may be written in terms of the functions $P(n, \theta, p)$ defined in equation (1) as follows:

$$\begin{aligned} P_1(p) &= \sum_{N_i=0}^{n_i} \binom{n_i}{N_i} p^{N_i}(1-p)^{n_i - N_i} \sum_{N_e = \theta + \phi N_i}^{n_e} \binom{n_e}{N_e} p^{N_e}(1-p)^{n_e - N_e} \\ &= \sum_{N_i=0}^{n_i} \binom{n_i}{N_i} p^{N_i}(1-p)^{n_i - N_i} P(n_e, \theta + \phi N_i, p). \end{aligned} \tag{17}$$

When n_e and θ are both large in such a way that we can use equation (2) to approximate to P, we get

$$P_1(p) \doteqdot \sum_{N_i=0}^{n_i} \binom{n_i}{N_i} p^{N_i}(1-p)^{n_i - N_i} S((\theta + \phi N_i)n_e^{-1}, p). \tag{18}$$

This means that $P_1(p)$ has various approximate functional forms over the various ranges delimited by the points at which the functions S take their steps. Thus:

$$
\begin{aligned}
&P_1(p) \doteqdot 0, \quad 0 \leqslant p < \theta/n_e,\\
&P_1(p) \doteqdot (1-p)^{n_i}, \quad \theta/n_e < p < (\theta+\phi)/n_e,\\
&P_1(p) \doteqdot (1-p)^{n_i}+n_i p(1-p)^{n_i-1}, \quad (\theta+\phi)/n_e < p < (\theta+2\phi)/n_e,\\
&\cdot \quad \cdot \quad \cdot \quad \cdot \quad \cdot \quad \cdot\\
&P_1(p) \doteqdot 1-p^{n_i}, \quad (\theta+n_i\phi-\phi)/n_e < p < (\theta+n_i\phi)/n_e,\\
&P_1(p) \doteqdot 1, \quad (\theta+n_i\phi)/n_e < p.
\end{aligned}
\tag{19}
$$

Of course we are only interested in the range $0 \leqslant p \leqslant 1$, so not all of the above forms for $P_1(p)$ necessarily occur in any one case. In particular, $P_1(1) = 0$ if $n_e < (\theta+n_i\phi)$ and $P_1(1) = 1$ if $n_e \geqslant (\theta+n_i\phi)$. In the very special case $n_i = 1$, we see that

$$P_1(p) \doteqdot 1-p \tag{20}$$

in the range $\theta/n_e < p < (\theta+\phi)/n_e$. This is illustrated in Fig. 5.5, where we

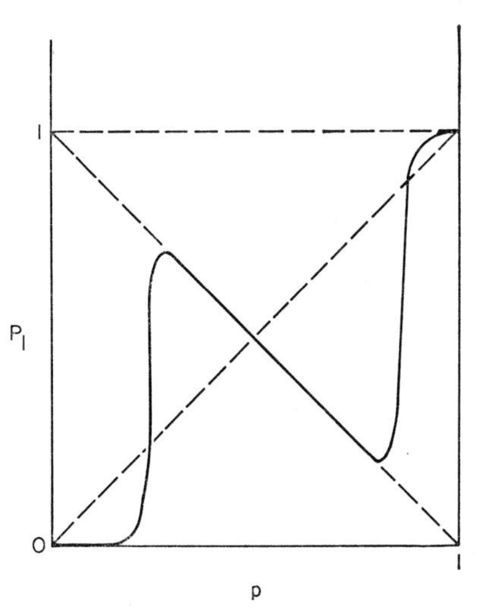

FIG. 5.5. Approximate common functional form of $P_1(p)$ when $n_i = 1$ (see text).

see that there are five values for which $P_1(p) = p$, namely $p \doteqdot 0$, θ/n_e, $\frac{1}{2}$, $(\theta+\phi)/n_e$, 1. We show in Section 5.3 that there is reason to expect the point $p \doteqdot \frac{1}{2}$ to be stable as well as $p = 0$ and $p = 1$, and stability may well be generally associated with a negative value for $P_1'(p)$ at the stationary point $P_1(p) = p$.

Such networks can also behave in a manner qualitatively similar to purely excitatory ones as we now see. Let ϕ be small so that the steps of equation (19) become very close together but keep n_e, n_i and θ large in proportion by writing them as $n_e = \mu_e n$, $n_i = \mu_i n$, $\theta = tn$, where μ_e, μ_i and t are fixed and we contemplate letting $n \to \infty$. Then we can again use de Moivre's theorem for both the binomial constituents of formula (16) and readily find

$$P_1(p) \doteqdot 0, \quad \text{when } p < t/(\mu_e - \phi\mu_i)$$
$$P_1(p) \doteqdot 1, \quad \text{when } p > t/(\mu_e - \phi\mu_i)$$

which closely matches the behavior for the excitatory networks shown in equation (2).

5.2.2. Comparison with Experiment

If we suppose that a random network settles down to one of the presumed stable stationary values for p, i.e. a value for which $P_1(p) = p$ and $P_1'(p) < 0$, there must result certain implicit numerical relations between the neuronal parameters n_e, n_i, θ and ϕ on the one hand and the network activity p on the other. Is there any way in which we could apply these relations to the brain and test them? Such a comparison would make the mathematical assumption that if we first replaced brain neurones with McCulloch–Pitts neurones and then randomized the connections, the relation between p and (n_e, n_i, θ, ϕ) would not be enormously altered. Frankly it is impossible to tell if it would or not, but it is interesting to make the comparison nevertheless to see how it works out.

We know that the mean firing rate for the brain is probably very low (1–10/second or less) compared with the rate at which neurones could fire. Equation (19) in general admits many solutions to $P_1(p) = p$ but if we suppose the brain typically operates with the lowest non-zero firing rate it can achieve having $P_1'(p) < 0$ we require that

$$(1-p)^{n_i} = p, \tag{21}$$

and that this value of p must satisfy

$$\theta/n_e < p < (\theta+\phi)/n_e.$$

The solution of equation (21) which is approximately correct for large n_i is

$$p \doteqdot n_i^{-1} \ln n_i \tag{22}$$

as is easily verified by taking logarithms of equation (21). Thus by taking n_i large, p becomes as small as we wish. As p must satisfy $p > \theta/n_e$, this means that about the lowest stable non-zero firing rate available is $p \approx \theta/n_e$ (this should be corrected for low θ as discussed in relation to Theorem 7). If we take $\theta \approx 150$–300 and $n_e \approx 30{,}000$ this gives $p \approx 0.005$–0.01. This should presumably be converted into a firing rate f/second by dividing by the McCulloch–Pitts time $\tau(\approx 5 \times 10^{-3}$ seconds) to give $f \approx 1$–2. Note that the real-time neurone expression (Section 5.1.5) is $\varepsilon\theta/n_e$ and gives a similar result.

This calculated value for f is in very happy agreement with the observed rates of 1–10/sec. It depends, of course, on an almost ridiculous number of assumptions and from one point of view may be rightly condemned. However, I think it is not devoid of interest that we get this order of magnitude agreement so easily between the experimental quantities n_e, θ and f.

Another way of looking at this matter is as follows. Take any network W composed of logical neurones (without habituation) and form a new network W_e by replacing every inhibitory connection with an excitatory one so that, for example, a neurone which had n_e excitatory and n_i inhibitory inputs in W gets n_e+n_i excitatory inputs in W_e. Then at time $p\tau$ suppose that a particular set of neurones in W is active and also just the corresponding set in W_e. Then at time $(p+1)\tau$ every neurone in W_e which corresponds to an active neurone in W at that time must be active (and also possibly others). This is because if, for a given neurone in W, $N_e-\phi N_i \geqslant \theta$ then

$$N_e+N_i \geqslant N_e \geqslant \theta+\phi N_i \geqslant \theta$$

which is the threshold condition for the corresponding neurone in W_e.

Hence if activity dies away in W_e, we can deduce that it also must in W. Conversely, if W has a self-maintaining activity, and we start W_e in the same initial state as we did W then we can deduce that the activity in W_e cannot die away to zero. As a consequence, it may even be quite accurately true to say that W cannot have a lasting level of firing which is below the threshold p_0 of that network which is obtained from W_e by converting it into a randomly connected network with the same total number of connections and θ equal to the mean of those in W_e. This would mean that the mean firing rate of the brain must at least satisfy $f \geqslant p_0/\tau$. We found $p_0 \simeq 0.005$–0.01 but the actual numerical value could be quite different because the value for θ is very uncertain.

One very interesting consequence of this inequality is that it suggests that the brain can only maintain a low level of firing activity if $\theta/(n_e+n_i)$ is low, which gives a possible significance to the high values for (n_e+n_i) which are observed (note, however, that this argument could be invalidated if the brain possesses large numbers of "pacemaker" cells having an intrinsic firing activity). A low firing level might be desirable for "economic" reasons based on the metabolic energy expended (for another possible reason related to memory see Griffith and Mahler, 1969; Griffith, 1970).

5.3. Differential-difference Equations

In the nervous system we are naturally particularly interested in the sequences of firing times of the nerve cells. However, the influence one nerve cell can have on another only takes place after a certain delay made up mainly of the axonal conduction time and the synaptic delay time. This means that it is to be expected that the mathematics of equations containing delay terms in them will often be of importance in the theory. Such equations are called differential–difference equations and they have properties which are peculiar in some respects. We shall give an introduction to them here, referring the reader to Bellman and Cooke (1963) for further reading on this topic.

We consider one specific example, which enables us to bring out a number of general points. In equation (20) we saw that the function $1-p$ was a good approximation to $P_1(p)$ when $n_i = 1$ and $\theta/n_e < p < (\theta+\phi)/n_e$. By taking θ small and ϕ large we can make this range go from near $p = 0$ up to $p = 1$. We shall use this simple approximate function here in order to attempt to discuss the stability of the network near the activity represented by $p = \frac{1}{2}$. We shall remove the condition of quantization of time but retain a delay between the time of firing of one cell and its influence upon synaptically-connected neighbors. If we still use the symbol p for a measure of mean activity of the network at time t, we shall now write it $p(t)$ and let it depend upon activity at previous times up to $t-\delta$. Specifically we shall consider the equation

$$p(t) = \int_0^\infty P(p(t-\delta-x))j(x)\,dx, \tag{23}$$

where

$$j(x) = \varepsilon\, e^{-\varepsilon x},$$

represents the relative weighting of various times in the past meaning, in effect, that the delay has a range of values from δ on up. Equation (23) is a generalization of the corresponding equation with $\delta = 0$ which was discussed previously by Griffith (1963a, pp. 306–8). It is not suggested, however, that it can be fully and rigorously justified for a network. We have chosen it rather because it has some physical plausibility and also because it has a transparent and tractable mathematical theory which we can use to illustrate the possible relevance of differential–difference equations.

Using formula (20) we can now convert equation (23) to differential rather than integral form as follows (dashes represent differentiation with respect to time):

$$\begin{aligned} p'(t) &= \int_0^\infty P'(p(t-\delta-x))j(x)\,dx \\ &= -P(p(t-\delta-x)j(x)\Big|_0^\infty - \int_0^\infty P(p(t-\delta-x)).\varepsilon j(x)\,dx \\ &= P(p(t-\delta))j(0) - \varepsilon \int_0^\infty P(p(t-\delta-x))j(x)\,dx \\ &= \varepsilon - \varepsilon p(t-\delta) - \varepsilon p(t). \end{aligned} \tag{24}$$

This is a differential equation because of the presence of $p'(t)$ and a difference equation because of the delay term $p(t-\delta)$. It is a differential–difference equation. If it has any stationary solution $p(t) = c$ (constant), then the constant c must satisfy

$$0 = \varepsilon - 2\varepsilon c,$$

and so $c = \frac{1}{2}$. We shall investigate the stability of this stationary solution by writing $p(t) = \frac{1}{2} + \xi(t)$, whence $\xi(t)$ satisfies the homogeneous equation

$$\xi'(t) + \varepsilon\xi(t) + \varepsilon\xi(t-\delta) = 0. \tag{25}$$

First let us ask what would happen if $\delta = 0$? Then equation (25) becomes

$$\xi'(t) + 2\varepsilon\xi(t) = 0$$

with the solution

$$\xi(t) = A\, e^{-2\varepsilon t}.$$

Hence as $t \to \infty$, so $\xi(t) \to 0$, $p(t) \to \frac{1}{2}$. The stationary solution is stable.

How do we proceed if $\delta \neq 0$? Again we wish if possible to show that any $\xi(t) \to 0$. Here we meet the most peculiar feature of these equations. Although equation (25) is first-order and linear we cannot give a general solution in terms of one arbitrary constant alone (such as A above). This is because we are free to assign $\xi(t)$ arbitrarily in a whole interval, $0 \leqslant t < \delta$ say, and then calculate it progressively from equation (25). It is a consequence of this that, as we shall see, equation (25) possesses an infinite number of linearly independent solutions.

In spite of this we can still adopt what is the most straightforward generalization of the procedure used for linear differential equations with constant coefficients. Namely, we search for special solutions of the type $\xi(t) = e^{st}$, where s is a constant. Having found these we can rely on the theorem (Bellman and Cooke, 1963, chapter 4) that every solution can be expressed as a linear combination, possibly with an infinite number of terms, of these special solutions. Therefore if we can show that every such *special* solution tends to zero as $t \to \infty$, we can deduce that *every* solution of equation (25) must do so and therefore that the stationary solution $p(t) = \frac{1}{2}$ is stable.

Now put $\xi(t) = e^{st}$ in equation (25). We have

$$s\, e^{st} + \varepsilon\, e^{st} + \varepsilon\, e^{s(t-\delta)} = 0,$$

which means that s must satisfy

$$s + \varepsilon + \varepsilon\, e^{-s\delta} = 0. \tag{26}$$

This is simplified by setting $z = s\delta$, $\beta = \varepsilon\delta$, whence

$$z\, e^{z} + \beta\, e^{z} + \beta = 0. \tag{27}$$

We need to solve this for $z = s\delta$ in terms of $\beta = \varepsilon\delta$. Furthermore in order to prove that $e^{st} \to 0$ as $t \to \infty$, we must show that the real part of s and hence of z is negative. First suppose z is real, then $z = -\beta - \beta\, e^{-z}$ and as $\beta > 0$ it follows that z must be negative, which deals with this case.

Next, let $z = x+iy$. Then $\xi(t) \to 0$ as $t \to \infty$ if and only if $x < 0$. Take the imaginary part of equation (27), which is

$$x\, e^x \sin y + y\, e^x \cos y + \beta\, e^x \sin y = 0$$

or

$$x = -y \cot y - \beta. \tag{28}$$

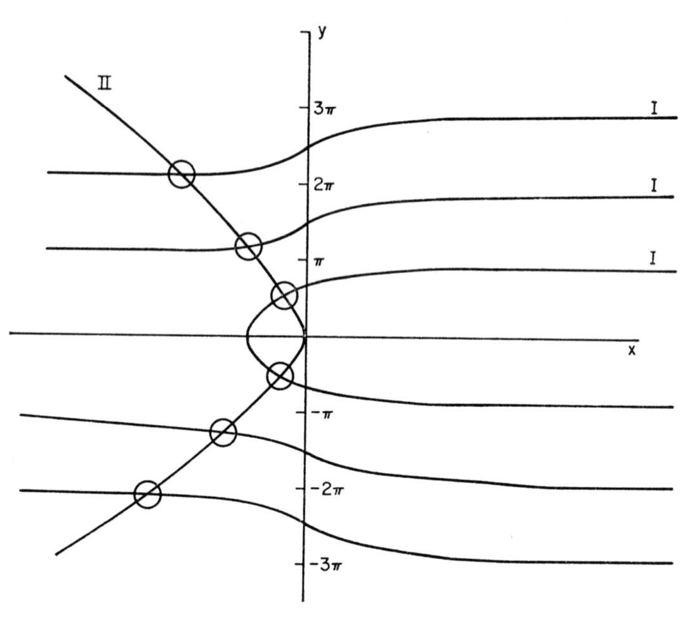

FIG. 5.6. Schematic plots of the curves in the complex plane generated by (I) equation (28) and (II) equation (29). The intersections (circled) of these curves are seen to form two infinite families and yield the solutions of equation (27).

Then take the square of the modulus of $(z+\beta)e^z = -\beta$ which gives

$$\{(x+\beta)^2 + y^2\}e^{2x} = \beta^2$$

or

$$y^2 = \beta^2\, e^{-2x} - (x+\beta)^2. \tag{29}$$

The allowed values of z lie at the intersections of the two curves (28) and (29) when they are plotted in the (x, y) plane. It is easy to see that they must all have $x < 0$ because if $x > 0$ we deduce from equation (29) that

$$y^2 < \beta^2 - (x+\beta)^2 < 0$$

which is impossible, and if $x = 0$ then $y = 0$. But in that case, from equation (28): $-y \cot y - \beta = -1 - \beta \neq x$. Hence all the special solutions $e^{st} \to 0$ as $t \to \infty$ and so therefore do all solutions, because they can always be expanded as

$$\xi(t) = \sum A(s)\, e^{st}. \tag{30}$$

The actual curves which determine the allowed values of $z = x + iy$ are sketched in Fig. 5.6, showing that there are in fact an infinite number of such values.

CHAPTER 6

Information Theory and Memory*

6.1. Introduction

In order to clarify the nature of a question about the brain, it is sometimes useful to contemplate asking the analogous question about a digital computer. The reason is that in the latter case we can examine in detail the structure both of the question and the answer as it applies to a well-defined and well-known structure. This can help us in formulating the question, as it applies to a brain, even though the answers in the two cases may often be completely different in detail. If we adopt such a procedure, we are not saying that a digital computer is necessarily at all like a brain in its construction but merely that we may often ask the same questions about it. In the case of memory, I think this helps us to bring out clearly the existence of a very fundamental quantitative problem.

A centrally important parameter to the user of a computer is its store size, or store capacity. Loosely speaking, this can be given as the number of different numbers it can hold but is best defined as the number n of bits (see Section 6.2.1) that can be stored in it. n is the same irrespective of the way in which the machine is being used, providing that one allows for the fact that programs take up storage space, and is an intrinsic constant of a particular machine. Of course, in practice we usually know its value because we know how much store the manufacturer put into the machine. But if we did not, the question "How many bits is the storage capacity of the computer?" would remain a question with a clear-cut answer, namely n. We could tackle this question by having a theory about the nature of the storage elements, i.e. that they were the ferrite rings each of which could store one bit, and then calculating n from direct observation of the hardware of the computer (in practice, of course, this particular theory would probably lead to an underestimate of n). Or, we might ask someone who knew how to program the machine in some high-level language, but did not know n, to see how many numbers he could get the machine to read in and store and then, subsequently, read out again correctly. This method might give many

* This chapter is largely an updating of Griffith (1965b).

different estimates of the capacity, none of which could exceed the true value but some of which, hopefully, might be near to it. Furthermore, if, by this second method, we found a much higher value for n than we did with the first, then this would furnish clear evidence that the theory we used in the first method was incorrect or at least incomplete. In neither method do we need to know anything at all about the detailed construction of the machine, considered as an integrated system, nor do we need to know how the programming language is implemented in the machine nor even, providing numbers are somehow read into it and read out again, do we need to know how to program it ourselves.

We need not pursue this analogy much further. However, it suggests to us that we ask whether a brain has a definite memory capacity, whether we can obtain theoretical estimates for this capacity in terms of definite hypotheses about the chemical or physical nature of the storage and whether we can obtain experimental estimates of it from observations of what animals can learn and reproduce and, if so, finally how these two kinds of estimate compare with one another. This is the object of the present chapter. I think that by far the most penetrating approach to memory capacity in animals, as in computers, is through information theory. This is the way we shall proceed, starting in Section 6.2 after a few more preliminary remarks.

In a computer there are usually several kinds of store differing in capacity and speed of access. Although at any one time, some items remembered by the brain are more easily and rapidly accessible than others, the possible distinction between different kinds of memory which has been most clearly recognized relates to permanence. Based on the pattern of amnesia arising from concussion (Russell, 1948, 1959), electroconvulsive therapy (see Deutsch, 1962) and, most recently, the injection of compounds which interfere with protein synthesis (Agranoff, Davis and Brink, 1966; Barondes and Cohen, 1966), it appears likely that a memory of an event is stored first in a relatively vulnerable or labile form (short term memory) and appears later in a form which may last for years or for a lifetime (long term memory), although even in the latter form there may be a continual further strengthening (see Russell, 1959, p. 16 and comment in Griffith, 1970). Our discussion of memory capacity should be thought of as relating to long term memory and, because of the difficulty of obtaining some of the necessary data about learning in other animals, mainly to man. Furthermore, by singling out the problem of memory capacity we are not implying that we think it is the only important problem. The question of the detailed mechanics of the storage and retrieval of the data in memory is, of course, of enormous interest and importance but is largely outside the scope of this book although it will be discussed briefly in Sections 6.5.1 and 6.5.3.

6.2. Information Theory

6.2.1. Capacity of an Information Store

I shall give an introduction to information theory which, although it follows traditional treatments, is nevertheless oriented clearly toward the biological situation of storage of information. Information storage is a term which is used in many ways in the common language, even by scientists, and information theory provides a particular way of assigning numerical values to it. Information theory is usually regarded as a statistical theory. However, the conditions which need to be satisfied in such an approach are more restrictive than is acceptable in many biological situations, and therefore we present first that part of the theory which can be treated without assumptions of a statistical nature. (For fuller treatments of information theory, see Shannon and Weaver, 1949; Khinchin, 1957; Brillouin, 1962; Gallager, 1968.)

We lead into the idea of capacity with the example of a valid voting return in an election involving two candidates. Such a return will have been filled in in one of two significantly distinct ways. It records which choice, out of two possible choices, the voter has made. It could be called a two-choice store of information. Similarly, for n candidates, with the possibility of voting for one only, the voting paper forms an n-choice store.

Alternatively, we may think of an n-choice store as something which can be set into one of n possible states. If each of these states is connected with some possible event, then we may set the store in the state corresponding to one of these events as a mnemonic or memory for the event corresponding to that state. A typical example here would be the notices often hung in shop windows which have two possible states, one presenting to the outside world the word OPEN and the other the word CLOSED.

Evidently the capacity of such a finite information store, sensibly, can only be either the number n of states, or at least some function $f(n)$ which increases with n. A suitable form for $f(n)$ is suggested by a consideration of what happens when we put two stores together to form a joint, larger store. If the two constitutent stores have, respectively, n_1 and n_2 states, then the combined store has $n_1 n_2$ different states. It is natural to hope that we could define capacity in such a way that the size of the larger store should be the sum of the sizes of its constituents. Thus we require

$$f(n_1 n_2) = f(n_1)+f(n_2). \tag{1}$$

If we assume that $f(n)$ also satisfies $f(n_1) > f(n_2)$ whenever the integers (n_1, n_2) satisfy $n_1 > n_2$, which is a second natural requirement, it is not difficult to show that the only possible function is

$$f(n) = a \log_b n \tag{2}$$

for any constants $a > 0, b > 1$. It is usual, although not universal, to choose $a = 1$ and $b = 2$. Then

$$f(n) = \log_2 n \tag{3}$$

and we shall always use this choice and drop the subscript 2 in equation (3). As usual, we shall use ln to denote logarithms with base e.

Now consider a store which consists of x switches in a line, each of which may be put up or down. Then the store has 2^x states and hence a capacity of x. On the other hand, if we represent "up" by the digit "1" and "down" by the digit "0", any state of the store gets represented by a number in the scale of two, the number having x digits. Conversely, each such number corresponds to a state of the store. Because of this relation between a store of capacity x and numbers with x binary digits, it is customary to say that any store with n distinct states has an information capacity of $\log n$ bits of information. Here bit is the conventional abbreviation for "binary digit".

Important examples of stores are easy to find. A modern digital computer operates entirely with elements, such as the magnetic ferrite rings of the core store mentioned earlier, each of which can be set in one of two alternative states. In the case of the ferrite rings this is because they can be magnetized in either of the two possible directions around the ring. Hence one may specify the storage capacity of the computer easily and definitely as a number of bits of information. In this case each bit is usually stored separately in a physically distinct position. However, this is obviously not necessary. The genetic material, DNA, which we shall discuss in Section 6.5.1 is often called a store of genetic information. It consists of linear polymers, in various possible orders, of four constituent monomers. Hence each position in the sequence of monomers admits of four chemical possibilities and can therefore store $\log 4 = 2$ bits of information. So a polymer of n monomers can store $2n$ bits.

One thing to note about the numbers involved is that even a fairly small information capacity, say 80, can correspond to the store having a quite astronomical number of different distinct states. In this case it is 2^{80} and if one tried to set the store sequentially into each of them, one after another, taking 1 microsecond for each state then the total time taken would be about 30 000 000 000 years which is comparable with the probable "age" of the universe. In the literature, people have often been unduly impressed with the numbers they have obtained after attempts to calculate things like the number of different possible patterns of neural activity, etc.

Now let us conclude this subsection by noting what I think is the most important thing about this preliminary definition and approach to information content. It is just that it contains no mention of probabilities. We shall see shortly that the statistical theory of communication leads naturally to the same definition of information capacity. But that approach is more restricted, for it has nothing to say about situations for which probability has no meaning. We may use a traditional example here, popular with philosophers. Let us, at midnight, place a switch in the position "off". Then we arrange that, when the sun rises in the morning, it is switched to "on". If we consult the switch at midday it tells us whether the sun rose that morning. It is acting as a store with a 1-bit capacity. However, the majority of statisticians would, I believe, consider that one cannot sensibly talk about the probability of this particular event. It would seem that, at the very least, it is not obvious that all the events in an animal's life can usefully be treated in probabilistic terms, although presumably some may. For this reason it is useful to have a definition of information capacity which transcends the probabilistic situation.

6.2.2. The Statistical Theory of Information

The statistical theory is only concerned with events which have definite probabilities, or frequencies. In the simplest non-trivial case we would have two possible events. For example, suppose an experiment is performed which has the possible outcome A_1 with probability p and outcome A_2 with probability $1-p$. If $p = 0$ or 1, we are essentially certain of the outcome and learn nothing by performing the experiment. If $p = \frac{1}{2}$, we are entirely uncertain beforehand and the experiment is very informative, If, say, $p = 0.9$ we are "fairly sure" of the outcome but not certain. In the statistical theory we define a quantity H which gives a useful measure of the degree of uncertainty before the experiment is performed. After it is performed, no uncertainty remains, and therefore H is said to be the information acquired, on average, when the experiment is performed.

As we shall see in a moment, there are a number of conditions one would expect H to satisfy. If these are satisfied, it turns out that the only acceptable definition of H for an experiment with n possible outcomes having respective probabilities $p_1, \ldots, p_n$ is

$$\mathrm{H} = -a \sum_{i=1}^{n} p_i \log_b p_i \tag{4}$$

and we shall take $a = 1$, $b = 2$ as before to give

$$\mathrm{H} = -\sum_{i=1}^{n} p_i \log p_i. \tag{5}$$

When $n = 2$,

$$\mathrm{H} = \mathrm{H}(p) = -p \log p - (1-p) \log (1-p) \tag{6}$$

and this is plotted as a function of p in Fig. 6.1.

We now check that H has the right sort of properties. We have $\mathrm{H}(0) = \mathrm{H}(1) = 0$, and so we acquire no information when we are certain of the outcome beforehand. Similarly, we acquire the maximum amount when $p = \frac{1}{2}$, as would be expected. This latter property may be generalized; that is, for any n, the function H of equation (5) attains a unique maximum

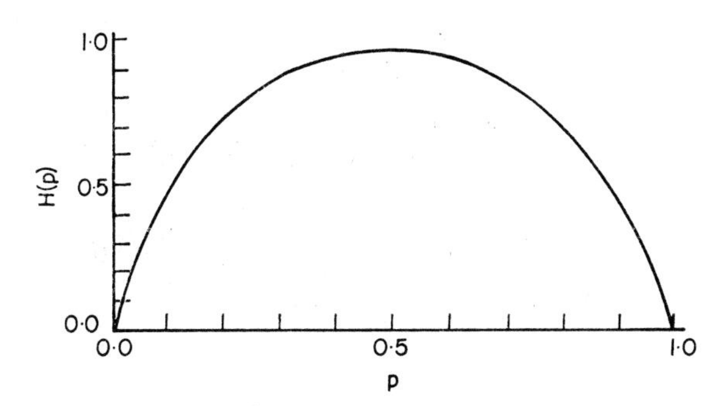

FIG. 6.1. $H(p)$ as a function of p for an experiment with two possible outcomes.

value when every $p_i = n^{-1}$ (Khinchin, 1957). When p_i has these values, we find

$$H(n^{-1}, n^{-1}, \ldots, n^{-1}) = \log n, \tag{7}$$

which is to be compared with equation (3).

Now consider the situation when we have an experiment A followed by another experiment B. Suppose them to be independent, with respective independent probabilities $p_1, \ldots, p_n$ and $s_1, \ldots, s_m$. Then the joint experiment has nm possible outcomes with probabilities $p_i s_j$, and hence

$$\begin{aligned} H(AB) &= -\sum_{i=1}^{n}\sum_{j=1}^{m} p_i s_j \log(p_i s_j) \\ &= -\sum_{i=1}^{n}\sum_{j=1}^{m} p_i s_j(\log p_i + \log s_j) \\ &= -\sum_{i=1}^{n} p_i \log p_i - \sum_{j=1}^{m} s_j \log s_j \\ &= H(A) + H(B). \end{aligned} \tag{8}$$

Thus the uncertainty of the joint experiment is the sum of the uncertainties of the two separate constituent experiments.

If A and B are not independent, this result cannot possibly hold. For example, the outcome of A might determine the outcome of B, in which case one would expect to find

$$H(AB) = H(A)$$

whatever the value of H(B). Let us write, then, π_{ij} for the probability of outcome j for experiment B, assuming the outcome i has already occurred for experiment A. Then the joint probabilities are now $p_i \pi_{ij}$. Because, for any i, there must be some outcome for experiment B, we have

$$\sum_{j=1}^{m} \pi_{ij} = 1, \quad (\text{all } i) \tag{9}$$

and

$$\begin{aligned} H(AB) &= -\sum_{i=1}^{n} \sum_{j=1}^{m} p_i \pi_{ij} \log (p_i \pi_{ij}) \\ &= -\sum_{i=1}^{n} \sum_{j=1}^{m} p_i \pi_{ij} (\log p_i + \log \pi_{ij}) \\ &= -\sum_{i=1}^{n} p_i \log p_i - \sum_{i=1}^{n} p_i \sum_{j=1}^{m} \pi_{ij} \log \pi_{ij} \\ &= H(A) + \sum_{i=1}^{n} p_i H_i(B), \end{aligned} \tag{10}$$

where

$$H_i(B) = -\sum_{j=1}^{m} \pi_{ij} \log \pi_{ij} \tag{11}$$

is the uncertainty about B after outcome i is known to have been obtained for A. The quantity

$$\sum_{i=1}^{n} p_i H_i(B)$$

is the mathematical expectation of H(B) in the situation that A has been performed first. That is, it is the average uncertainty in B remaining after the outcome of A is known. We write it $H_A(B)$, and then equation (10) becomes

$$H(AB) = H(A) + H_A(B) \tag{12}$$

which is intuitively very acceptable. In ordinary language, it may be read: The amount of uncertainty in the joint experiment A followed by B is equal to the sum of the uncertainty in the experiment A and the average uncertainty in B which remains after A has been performed.

The function $H = H(p_1, p_2, \ldots, p_n)$ also possesses two further properties, namely,

$$H \geqslant 0 \tag{13}$$

$$H(p_1, p_2, \ldots, p_{n-1}, 0) = H(p_1, p_2, \ldots, p_{n-1}). \tag{14}$$

Clearly, it is desirable that H should possess all the properties represented in equations (12)–(14) and also have its maximum for $p_i = 1/n$. It can be shown that the H of equation (4), with $a > 0$, is the only function of the p_i having all these properties (Khinchin, 1957). In other words, we have essentially no choice in our definition if we wish H to have reasonable properties. Then H represents the degree of uncertainty before the experiment is performed and therefore also the information obtained by performing the experiment.

6.2.3. Communication and Channel Capacity

Suppose now that our experiment A is performed by a first observer who communicates his result to a second observer. It is natural to say that he communicates a quantity of information H. In particular, if the second observer is replaced by a store which has n states and can therefore be set to record which of the n outcomes of experiment A has occurred, the store will then have an average content of information H. The store has capacity

$$C = \log n \tag{15}$$

and is storing an average amount H, where H satisfies

$$0 \leqslant H \leqslant \log n = C. \tag{16}$$

Thus, from the statistical viewpoint, the capacity of a store is the maximum amount of information which can be stored. However, unless each $p_i = 1/n$, the actual amount stored will be less than this.

The process of transfer of information from the first observer to the store gives an example of a communication channel. In statistical communication theory, such a transfer is regarded as having three components—a source, a channel, and a receiver (Shannon and Weaver, 1949). The source emits a series of symbols according to some probability regime. In our example, the probabilities of successive symbols are independent, but this assumption is not necessary. It is usual, however, to assume that the sequence forms a stationary and ergodic stochastic process (Khinchin, 1957). Essentially, this means that the probability regime is the same all along the sequence, although individual probabilities may depend on which symbols occurred earlier in the sequence. It also denies the existence either of periodicity or that the process might be composed of entirely independent subprocesses. The latter would occur, for example, if the source could emit four possible symbols $A_1, \ldots, A_4$ but in fact, with probability p, always (i.e. for all time) emitted A_1 or A_2 and, with probability $1-p$, always emitted A_3 or A_4.

The channel carries the symbols and the receiver receives them. In our example the receiver was the store. Clearly, the store is not allowed to register a symbol for long and would get reset each time a new symbol is transmitted. If we wish to have a store which acts as a permanent memory for all the symbols sent in, say, N transmissions in the channel, we need N replicas of our store. Hence the total necessary capacity of store is $N \log n$, and it can store NH bits of information. Evidently,

$$NH \leqslant N \log n \tag{17}$$

and the capacity of the store is generally greater than the information it actually holds.

This brings us to the idea of coding, by means of which, when

$$NH < N \log n \tag{18}$$

we may generally use a store of smaller capacity to store practically the same amount of information.

Before stating a theorem about this, let us consider a simple example. Suppose a source emits three symbols 1, 2, or 3 with probabilities $p_1 = 0.8$, $p_2 = p_3 = 0.1$. Then it emits information at a rate

$$\begin{aligned} H_e &= -0.8 \log 0.8 - 2 \times 0.1 \log 0.1 \\ &= 0.9217. \end{aligned} \tag{19}$$

The capacity of stores having two and three states are, respectively, log 2 = 1 and log 3 = 1.58. Obviously the second of these two stores will accurately record the transmitted symbols (with erasure before each new symbol, as discussed above). The first store has only two states, A_1 and A_2, say, and might be used in the following way. Set A_1 if symbol 1 is received and set A_2 if symbol 2 or 3 is received. Then we have, in the store,

$$\begin{aligned} p(A_1) &= 0.8; \qquad p(A_2) = 0.2 \\ H_s &= -0.8 \log 0.8 - 0.2 \log 0.2 \\ &= 0.7218. \end{aligned} \tag{20}$$

We have lost an amount of information

$$\Delta H = H_e - H_s = 0.1999 \tag{21}$$

and this is because we cannot tell by inspecting the store whether the symbol 2 or 3 was transmitted, in the case that either of them has been.

This gives a simple example of a code. However, we could use a different code and set A_1 when 1 or 2 is transmitted and set A_2 when 3 is transmitted. Then

$$\begin{aligned} p(A_1) &= 0.9; \qquad p(A_2) = 0.1 \\ H_s &= 0.4689; \qquad \Delta H = 0.4528. \end{aligned} \tag{22}$$

We have lost a much larger amount of information. Evidently our first code is a more efficient one.

Instead of concentrating on the store, we may suppose the channel has only two symbols it can transmit. Then the three emitted symbols must be coded in terms of the two, A_1 and A_2 of the channel. Finally, the channel transmits A_1 or A_2 to the receiver. The channel is now said to have a capacity of C = log 2 = 1. This capacity would be achieved if A_1 and A_2 have equal probabilities, for then 1 bit per symbol is transmitted. However, in our example, at most 0.7218 bit per symbol passes along the channel.

We can do better than this by treating the sequence of symbols as a sequence of half as many pairs of successive symbols and then coding for each of the pairs as in the following tabulation.

Send	A_1A_1	A_1A_2	A_2A_1	A_2A_2
When *i*, *j* is	11	12, 22, 23	13, 32, 33	21, 31
Total probability	0.64	0.10	0.10	0.16

We readily find, for the information content per symbol,

$$\begin{aligned} H_s &= \tfrac{1}{2}\{-0.64 \log 0.64 - 2 \times 0.1 \log 0.1 - 0.16 \log 0.16\} \\ &= 0.7496 \qquad (23) \\ \Delta H &= 0.1721. \end{aligned}$$

With coding three successive symbols at a time, we can achieve

$$H_s = 0.7896; \qquad \Delta H = 0.1321. \qquad (24)$$

The information loss ΔH clearly decreases significantly from equation (21) to equation (24). Does it tend to zero with increasing coding length?

The answer to this is "yes" and is a particular case of Shannon's theorem for information transmission in a noiseless channel. A channel is noiseless if it is possible to reconstruct the sequence sent into the channel from the sequence received at the other end. In our example, the channel was noiseless because A_1 is received when A_1 is transmitted and similarly for A_2. For such a channel, Shannon's theorem asserts that, if the information content H of the source is less than the capacity $C = \log n$ of the channel, then, by coding with sufficiently long sequences, the information loss ΔH may be made arbitrarily small. If $C < H$, then this is not possible and we must always have $\Delta H \geqslant H - C$, although it may be made arbitrarily close to $H - C$ with suitable coding (although see Khinchin, 1957, p. 119).

This concludes our introduction to information theory, and we reiterate that the definition of capacity of a store does not depend on statistical considerations. However, that capacity coincides with the definition of capacity in the statistical theory. The statistical theory is presumably most useful in discussing constantly recurring events in an animal, such as the firing of nerve cells, or in its environment. Rare events, occurring perhaps once in a lifetime, are more difficult to discuss in this way, especially because of the usual requirement in information theory of stationary stochastic processes. Because of its birth, maturation, and death, the interaction of an animal with its environment can hardly be called a stationary process.

6.3. Information Content of Nerve Cell Firing Sequences

Let us now consider a nerve cell as a source emitting information. Its output is an action potential moving down its axon and essentially reproducible from one firing to the next. Hence the information is carried in the sequence of times t_n, n integral, at which the cell fires. Can we assign an information content, at least if we know something about the statistics of this sequence? First, note that such a sequence of firings may be thought of in one of two equivalent ways. They may be regarded either as the firings with their times of occurrence, or as the intervening time intervals between successive firings. It turns out that the latter is a rather more convenient way for discussing information content.

For reasons which will be apparent shortly, we shall start by supposing that there are only a finite number of distinct intervals possible. Let us label them $\lambda_1, \ldots, \lambda_p$, with p finite. We shall discuss only the case in which the intervals occur independently with probabilities $\phi_1, \ldots, \phi_p$, respectively. This means that the probability of occurrence of a particular interval is independent of which intervals have occurred earlier. Such a process is often termed a renewal process in statistics (see the end of Section 4.2). First we consider the information content H_I per interval. This is given immediately by the formula

$$H_I = -\sum_{i=1}^{p} \phi_i \log \phi_i \quad \text{with} \sum_{i=1}^{p} \phi_i = 1. \tag{25}$$

The mean length of an interval is

$$m = \sum_{i=1}^{p} \phi_i \lambda_i. \tag{26}$$

Therefore the information content H per unit time is given by the ratio

$$H = \frac{H_I}{m} = -\frac{\sum \phi_i \log \phi_i}{\sum \phi_j \lambda_j}. \tag{27}$$

Clearly, the value of H depends upon the probabilities ϕ_i. The capacity for the process is defined as the maximum of H for all choices of ϕ_i, that is

$$C = \max_{(\phi)} H. \tag{28}$$

To determine C, we use Lagrange's method of undetermined multipliers (see Appendix), in which we need to satisfy $\delta H = 0$ subject to the condition $\delta \Sigma \phi_i = 0$. We have

$$\begin{aligned} \delta H = \delta\left(\frac{H_I}{m}\right) &= \frac{\delta H_I}{m} - \frac{H_I \delta m}{m^2} \\ &= -m^{-1} \sum (l + \log \phi_i)\, \delta\phi_i - (H/m) \sum \lambda_i\, \delta\phi_i. \end{aligned} \tag{29}$$

In equation (29), $l = \log_2 e$ because the logarithms occurring in equation (27) are to base 2 not to base e. We now set

$$\delta H - \mu\delta \sum \phi_i = 0$$

and take the coefficient of $\delta\phi_i$. This is

$$-m^{-1}(l+\log\phi_i)-(\mathrm{H}/m)\lambda_i-\mu = 0$$

or

$$l+\log\phi_i+\mathrm{H}\lambda_i+\mu m = 0. \tag{30}$$

We now determine μ by multiplying through by ϕ_i and summing over i. This gives

$$\sum(l\phi_i+\phi_i\log\phi_i+\mathrm{H}\lambda_i\phi_i+\mu m\phi_i) = 0$$

or

$$l-\mathrm{H}_I+\mathrm{H}m+\mu m = 0$$

$$\mu = -l/m. \tag{31}$$

Using this value of μ, equation (30) simplifies to

$$\log\phi_i+\mathrm{H}\lambda_i = 0$$

and so

$$\phi_i = 2^{-\mathrm{H}\lambda_i}. \tag{32}$$

H itself is determined from the normalization condition for the probabilities given in equation (25). In other words,

$$1 = \sum\phi_i = \sum 2^{-\mathrm{H}\lambda_i}. \tag{33}$$

The preceding method is closely related to those used in corresponding situations in statistical mechanics. Section 6.3 is based on a discussion given by Wall, Lettvin, McCulloch and Pitts (1956), following a partially incorrect treatment by MacKay and McCulloch (1952). See also Rapoport (1955). Another derivation of equation (33) was given by Shannon in his classic work on information theory (Shannon and Weaver, 1949). The connection with Shannon's result is most easily seen by setting $y = 2^{-\mathrm{H}}$, so that $\mathrm{H} = -\log y$. The equation (33) becomes

$$\sum y^{\lambda_i} = 1. \tag{34}$$

Equation (34) is given in Shannon and Weaver (1949, p. 8).

We have established our result for p finite; then H is finite. It is given by taking that value of y which is the smallest real positive root of equation (34). In case we let $p\to\infty$, then H may become infinite, as we show in a particular case in a moment.

We shall now apply our general formulae to some examples.

Example 1. Here we suppose that all the intervals are multiples of a unit τ. Obviously, the value of τ merely sets the time scale, and so to start with we shall put $\tau = 1$. The simplest case obtains when we take all multiples of τ up to a certain value, say p. In other words we set $\lambda_i = i$. In this case, equation (34) becomes

$$\sum_{i=1}^{p} y^i = 1 = \frac{y^{p+1}-y}{y-1}.$$

Hence y is a solution of

$$f(y) \equiv y^{p+1} - 2y + 1 = 0 \tag{35}$$

where we reject a spurious root $y = 1$. The solution we are looking for is that y which is real, positive, and as near zero as possible. If $p = 1$, we have $y = 1$ and $H = 0$. This is to be expected, because in this case there is only one distinct length of interval and therefore the information content is obviously zero. If $p > 1$, then by considering the values of $f(y)$ at $\frac{1}{2}$ and 1 and the sign of $f'(y)$, it is easy to see that there is just one real root and that it satisfies $\frac{1}{2} < y < 1$. Hence $H = -\log_2 y < 1$. It is also easily shown that, as p runs from 0 to ∞, so H runs from 0 to 1. The actual values may be easily computed, and a number of them are given in Table 6.1.

TABLE 6.1

Values of H, allowing intervals of length 1, 2, . . ., p

p	H	p	H
1	0	6	0.988
2	0.694	7	0.994
3	0.879	8	0.997
4	0.947	9	0.999
5	0.975	∞	1

The fact that $H = 1$ for $p = \infty$ is to be expected because this means that at each point where it is possible to have a firing we now have a completely free choice of having one or not. Hence there is 1 bit of information per point, i.e. 1 per interval.

If $\tau \neq 1$, the values are the same as before but are now H per interval of length τ. Thus if time is quantized in units of $1/n$ seconds with, therefore, a maximum rate of n per second, then $H = n$ per second when $p = \infty$. It is also evident from this, that, if we allow intervals unrestricted in length, the information carrying capacity is infinite.

Example 2. We now consider the case in which all intervals are a multiple of a unit τ but with a minimum possible length $n\tau$ for an interval. This example illustrates the situation when we have a refractory period $R = n\tau$. Again put $\tau = 1$ at first. Then we have

$$1 = \sum_{i=n}^{p} y^i = \sum_{i=0}^{p} y^i - \sum_{i=0}^{n-1} y^i = \frac{y^{p+1}-1}{y-1} - \frac{y^n - 1}{y-1}$$

which reduces to

$$f_p(y) \equiv y^{p+1} - y^n - y + 1 = 0. \tag{36}$$

where again there is one spurious root at $y = 1$. As $p \to \infty$, the value of

y in which we are interested tends to the solution of

$$f(y) \equiv y^n + y - 1 = 0. \tag{37}$$

We shall only consider this equation, which corresponds to the situation when all intervals of length $n\tau$ and over are allowed.

When $n = 1$, we have $y = \frac{1}{2}$ and $H = 1$ in accordance with our previous example. When $n \to \infty$, we find $y \to 1$ and $H \to 0$, as might be expected.

Let us now consider what happens when the refractory period $n\tau = R$ remains constant but the time units τ become increasingly small. This will be found by letting $n \to \infty$ subject to R remaining constant. The information content per second is now

$$\begin{aligned} H' &= \tau^{-1}H = -\tau^{-1}\log y \\ &= -(n/R)\log y. \end{aligned} \tag{38}$$

We shall obtain an asymptotic expression for H′. This is most conveniently done by setting $y = 1 - \varepsilon(n)$. For n large, ε will be small. Equation (37) now becomes

$$y^n = \varepsilon(n). \tag{39}$$

This is of the same form as equation (5.21) and has the approximate solution

$$\varepsilon(n) = \ln n/n. \tag{40}$$

Using equation (39), this gives

$$H = R^{-1}(\log n - \log \ln n). \tag{41}$$

The next stage in the approximation is

$$H' = R^{-1}\left(\log n - \log \ln n + \frac{\log \ln n}{\ln n}\right). \tag{42}$$

This shows that even when we include the refractory period, the information carrying capacity per second still tends to infinity as $n \to \infty$ ($\tau \to 0$). There is no reason to suppose that the firing sequences arising from nerve cells have their intervals restricted in length, apart from the restriction imposed by the existence of the refractory period. Because of this, the information content of a nerve cell firing sequence is arbitrarily large, depending on the accuracy of our recording technique. Does this mean that we must assign an infinite information capacity to an axon carrying action potentials from one cell to a synapse with another? From a highly theoretical point of view the answer may be yes. However, in practice we must realize that the effect on the postsynaptic cell, as seen by its firing times, cannot be used to reconstruct the exact times of arrival of the presynaptic impulses beyond a certain accuracy. This is partly because of the integrative action of the cell whereby, for example, owing to an inability to reach the threshold the influence of many input impulses is completely lost. It is also due to the smearing out in time of the PSP, as discussed in Section 2.2.4, and to the variable size of PSP which is expected to arise from a single impulse. These questions do not seem to have been examined in detail but must mean that the post-synaptic cell can only "interpret" the sequences of input impulses that it receives to a certain accuracy, τ say, of the order of a millisecond. One approximate way of introducing this effect is to say that, since no intervals differing by much less than τ are recognizably different, then we should take all intervals to be a multiple of τ in the mathematical theory. This gives a convenient model, and we shall call τ the relevance time of the sequence.

However, although $H' \to \infty$ as $n \to \infty$ it only does so very slowly. The capacity is much more sensitively dependent upon the refractory period R. As an illustration of the dependence of H' on n, we find that, per unit of R, $H' = 1$ when $n = 1$; $H' = 1.39$ when $n = 2$; and $H' = 7.60$ when $n = 1024$. Thus in going from $n = 1$ to $n = 1024$, we only increase the capacity by a factor of 7.6 rather than by the factor of 1024, which would hold if there were no refractory period. Some idea of the accuracy of the approximation given in equation (42) is obtained by noting that it gives $H' = 7.61$ when $n = 1024$.

One result of the insensitivity of H' to n is that, from the information carrying viewpoint, there would not seem to be much point in making τ very small compared with R. In fact, one might expect τ and R to be of the same order of magnitude, which appears to be the case experimentally.

Example 3. We now consider what we may call a Poisson quantized process. At each point $0, \tau, 2\tau, \ldots, n\tau, \ldots$, let us suppose there is a probability p of firing and that the probability of firing at one point is independent of that at all other points. Then the interval of length $n\tau$ has the probability

$$P(n) = p(1-p)^{n-1}. \tag{43}$$

It follows immediately from equation (43) that

$$\sum_{n=1}^{\infty} P(n) = 1. \tag{44}$$

The information content H per unit τ is

$$H(p) = -p \log p - (1-p) \log (1-p).$$

This has its maximum at $p = \frac{1}{2}$, when $H = 1$ and $P(n) = 2^{-n} = 2^{-nH}$, which agrees with our earlier results, including equation (33).

We show H for various values of p in Table 6.2 (see also Fig. 6.1). The appropriate value of p is obtained by multiplying the firing rate f per second by the relevance time τ in seconds to give the probability of firing per unit τ of time. Herz, Creutzfeldt and Fuster (1964), found a mean value of $f = 35.5$ for 13 cells in the optic tract of the cat. This would give $p = 0.0355$ if τ is

TABLE 6.2

Values of H for Poisson quantized process

p	0.5	0.1	0.01	0.001
H	1	0.4689	0.0807	0.0113

a millisecond, whereupon $H(p) = 0.22$. Assuming these cells to be typical and that the time sequences of firing are not too far from being Poisson, this gives 220 bits/sec/cell. If this estimate could be taken over to the optic tract of man, which has about 2×10^6 axons (Section 2.2.6), we should get 4.4×10^8 bits/sec input for the whole optic tract. The total number of input neurones in man is about 5.3×10^6 (Section 2.2.6) which with this value of $f = 35.5$ would then give 1.2×10^9 bits/sec. However, 35.5 may easily be a serious overestimate for the average firing rate of all sensory input fibres to the brain and if it were replaced with 10 (or 1) the above figures would be reduced by a factor of about 2.7 (or 20). So perhaps we may expect the total input per second to be something of the order of 10^8–10^9 bits in man.

Note that whatever reasonable value we use for f, the information carried per second must be considerably less than the theoretical upper limit of $1/\tau$, corresponding to $p = \frac{1}{2}$, which is 1000 bits/sec if τ is 1 msec. Nerve cells are normally capable of firing at rates up to about 1000/sec, and therefore it would be possible for them to make full use of their potential information carrying capacity of 1000 bits/sec. Evidently, considerations other than these simple information-theoretic ones are responsible (cf. Section 5.2.2).

6.4. Human Memory Capacity

6.4.1. Information Capacity of a Brain

It is not easy to define precisely the information capacity of a brain. However, we should certainly distinguish between instinctive and learned behavior or knowledge. The former is presumably written into the brain, largely deriving from inherited instructions stored in coded form in the genetic material, as neuronal thresholds, patterns of connection and similar things. The latter needs a store and if the store has capacity C, this means it can be in any one of 2^C distinct states. We shall not concern ourselves with the first kind of knowledge although, obviously, a considerable amount of information capacity in DNA may need to be used for it and this also raises interesting questions. The second kind is not nearly as easy to distinguish from the first as might appear at first sight.

For example, suppose an animal has a highly complicated instinctive behavior pattern which is not completely specified at birth. Perhaps the value of a parameter needs to be set at one of x various possible values. Then unless this is realized, the resultant variability of the observed behavior may be taken to mean that the whole of that behavior has been learned. It might then take an external observer many thousands of bits to specify it in detail and he might erroneously suppose the animal can learn and store that number of bits rather than the very few bits which $\log x$ would be.

This consideration is not merely a logical possibility, but is almost certainly very important in practice. Interesting examples occur in the learning of song by birds. Thorpe (1961, especially Chapter 5) summarizes a lot of work in this field. In his own work on chaffinches, he finds that a bird brought up isolated from other birds, and unable to hear them, will nevertheless produce a song but of a much simplified nature. This song is, however, clearly related to the normal song. In other words, the bird sings something, but does not get it quite right unless it has the opportunity to listen to other birds and thereby "set a few parameters correctly". The word "correctly", here, means correctly relative to the songs of its fellows. Another example is given by the phenomenon of "imprinting" which is common in many if not all higher animals. Here the animal has a largely instinctive pattern of behavior toward its mother. However, it has to learn which is its mother and does this during a short and usually fairly well-defined period of early childhood. That it really learns this is shown by the fact that it can be made to accept a range of alternative "mothers". The ugly duckling was persuaded to accept a duck as its mother, but the modern experimenter can make it accept many other alternatives, even entirely non-living ones. The existence of imprinting is less certain for humans although it is well known for monkeys, but it is at least likely. Certainly there is a great range of similar phenomena occurring widely in the animal kingdom, including humans (Scott, 1962).

Another point to bear in mind is that abilities which are present later in life, but not earlier, are not necessarily learned. They can only reasonably be termed "learned" if they can be modified or suppressed by past experience. This would not always seem to be realized; for example, it cannot be deduced that the eye-closure reflex in humans must be learned, because it is not present in the new-born child. It does seem to be learned (Riesen, 1947), but this fact is not in itself sufficient evidence for it.

A related matter is that an animal (or human) may store a fact which, from the human scientist's point of view, seems to be a typical member of a set S consisting of n elements. The scientist might then be tempted to say that $\log n$ bits of information were stored. However, the brain may be storing according to entirely different principles, according to which the fact is a typical member of a set S′ with n' elements. $\text{Log}\, n'$ may be considerably less, or more, than $\log n$. An example of this for a computer is that, if one tried straightforwardly to store the first million primes in a computer with a million-bit memory, there would not be room. Write, however, a program which replaces the number x by the xth prime p_x and there would be room for this program. You can "store" the million primes in this way if you recognize that they are primes, i.e. that they belong to a limited class with special properties. Similarly, when an animal learns a complicated task, it may actually store less or more information than you would calculate from a particular classification of the choices needed to define the task.

Of course there are experiments in which it is possible to state a clear lower limit to the information recorded. If one asks a human to learn a sequence of n letters, each of which is chosen at random, it is natural to say that, on average, he has stored at least $n \log 26 \doteqdot 4.7n$ bits of information. We can only say this "on average", because it is well known from psychological experiments that "meaningful" sequences are learned more easily than "meaningless" ones, and so, perhaps, there is a sense in which we can say that the sequences are coded so that the more frequent, meaningful ones take up a smaller storage space in the brain than the less frequent, meaningless ones. That, of course, is mere speculation but it illustrates a consideration which has to be taken into account in particular cases.

This leads us on to the question: Can we obtain a lower limit of the human information capacity by asking how many books a man can learn by heart? I do not know of any precise attempt along these lines; however, I suppose the total content of the "Encyclopaedia Britannica" would be roughly comparable to a lifetime's work and might at least give an order-of-magnitude estimate. I have been interested to find since first writing those words (Griffith, 1965b) that Turing (1950), who was responsible for the Turing machine concept in mathematical logic, had in an early discussion of human memory capacity already used the "Encyclopaedia Britannica" as a standard of comparison. The problem of the information content of written English has been often discussed (see, e.g. Shannon and Weaver, 1949). There is, of course, the fundamental problem of whether it should be calculated on the assumption that English is a stationary stochastic process. If one leaves this difficulty aside, there is still the problem of estimating the extent of sequential dependence. If there is none, we have $H \doteqdot \log 27 = 4.75$ bits per letter or space. If we take into account the letter frequencies, which are of course not all equal, we get $H \doteqdot 4.1$. Taking into account higher sequential dependence gets us a lower value for H. An estimate due to Shannon is of the order of 1 bit per character (quoted in Pierce, 1967). This gives us about 2×10^8 bits for the information content of the "Encyclopaedia Britannica" (the Bible or Homer are something of the order of 10^6 bits). Even such a figure is to be regarded with caution in both directions. The information content per letter may be less, although perhaps it is not likely to be much less, but also the brain may or may not have a very efficient coding system. If H is the information content of the "Encyclopaedia Britannica" and C is the capacity necessary for a brain to store it, then C > H, but it may be that C is much greater than H.

6.4.2. Numerical Values

We shall now discuss the magnitude of the information capacity of the human memory, and we shall admit at the start that we do not know what it is. However, various methods give orders of magnitude, as our discussion of the "Encyclopaedia Britannica" did, and we shall not try to be more precise than that (for other discussions of this matter, see Turing, 1950; Von Neumann, 1958; Wooldridge, 1963; Griffith, 1967d, 1970). If C is the capacity, we shall express our results to the nearest integer as $\mu = \log_{10} C$. Our estimate from books was then, $\mu \simeq 8$ or, perhaps better, $\mu \geqslant 8$.

The next method may be called the psychological method. Here you measure the rate per second at which information of the type of words, patterns or pictures can be stored. Whether they are stored is tested by whether they can be reproduced at the time. Such methods are said to give of the order of 10–50 bits/sec (Quastler, 1956, 1965) although there is the usual problem of precisely how much information is presented in such cases. This is then grossed up over a lifetime. A lifetime has $1\text{–}3 \times 10^9$ sec and hence $\mu \doteqdot 10\text{–}11$. Note, however, that "forgetting" is ignored in this calculation, so if one believes the other premises of it, one would presumably write $\mu < 10\text{–}11$.

The next methods are based on physiological considerations. The largest input is via the visual tract, which has been estimated to give about 38% of the total, as measured in terms of number of fibers (Bruesch and Arey, 1942, and Section 2.2.6). In Section 6.3 we saw that the total input may be of the order of $10^8\text{–}10^9$ bits/sec, so if all of this is stored we get $\mu \doteqdot 17\text{–}18$ in thirty years. It seems extremely unlikely that a detailed record of the time sequences of all action potentials in all input fibers is in fact stored, so this calculation is unlikely to give a realistic estimate of human memory capacity.

With the exception of our first estimate, these values are based on grossing up the most favorable cases and assuming no forgetting. The assumption of no forgetting is usually defended by quoting rather spectacular examples of recall from the distant past, often under the influence of hypnotism or direct electrical stimulation of the cortex. All that such examples, and the reliable ones seem few in number, prove is that detailed accurate memory at the psychological level is possible over long periods, not that it always occurs. It is obviously unrealistic to be dogmatic about this matter at the present time, but my personal perusal of the psychological literature on learning with its normal concomitant, forgetting, makes me feel it is much more likely that these spectacular examples are atypical and misleading (see also Griffith, 1970).

My personal view is that, with our present limited knowledge, a reasonable estimate of μ at the present time is 9–10. Further, it remains to be proved that $\mu > 8$. I think there is no evidence that $\mu > 11$.

6.5. Discussion About the Nature of Memory

6.5.1. The Laying Down of Long-term Memory

The primary input to the brain is in the form of action potentials arriving along the axons of sensory nerves. This ultimately induces, or modifies, firing activity in the neurones of the brain itself. How does this activity take the first step towards being transformed into anything which is sufficiently enduring to merit the name of memory? We can do little here except to underline the importance of discovering the nature of this primary transformation, and to mention one or two possibilities, for the very good reason that the answer to this question is unknown.

One way to sidestep this problem, and to answer it at the same time, is to hypothesize that maintained self-re-exciting firing activity can itself serve as a permanent memory. We already saw in Section 3.1.2 that this is possible in principle and this type of dynamic memory theory has been discussed for quite a long time (see Burns, 1958, pp. 24–33). Such a mechanism for memory would have a close analogy with the mercury delay lines which were used in early computers and did in fact involve activity being continually cycled. If the brain were made of logical neurones, it is clear from the examples given in Section 3.1.2 that a memory capacity of up to one bit per neurone could be achieved, i.e. probably something of the order of 5×10^9 bits. In principle one could obviously improve on this with real neurones but it would seem unlikely that one could do much better in practice, with memories maintained over any great length of time. Hence if one proposed that all long-term memory were of this form, one would probably feel that the capacity was a little low in relation to our discussion in Section 6.4.2, especially as a nerve cell used for memory in this manner can hardly be used for much else. Therefore not all nerve cells are likely to be so used, which would lower the available memory capacity. Another line of objection to such a theory of long-term memory is the apparent implication of vulnerability of the stored information. One might expect to be able to clear the store, or at least completely disrupt it, by the general and violent outburst of cell firing associated either with electroconvulsive therapy or with grand mal epilepsy. That is, one might expect to erase completely all memory, which is not observed. Experiments have also been performed in which animals have been cooled in order to try to abolish electrical activity, which again appears to have little effect upon memory (Andjus, Knöpfelmacher, Russell and Smith, 1956). All in all, it is probably fair to say that this theory of long-term memory may be regarded as very improbable.

If we reject the dynamic self-re-excitation type of memory theory, then the most natural possibility to consider is that memory consists of some alteration in the structure of the network of nerve cells. This could occur through the formation of new synapses, or the destruction or alteration in strength of existing ones. One rather special way in which destruction of synapses could occur is through the death of cells (see Burns, 1958). Although it may seem more likely that this occurs fairly well at random and carries with it a loss of memory it cannot be said that the alternative possibility of a controlled destruction serving a function in memory has been completely excluded. The idea that alteration in connectivity is at the basis of memory has existed in one form or another for a very long time and is probably still the most plausible theory (it was discussed by Tanzi in 1893 and Cajal in 1895; see Cajal, 1952, vol. 2, pp. 886–7).

A related possibility is that the thresholds of neurones might be capable of lasting alteration and that this might serve as a basis for long-term memory (Shimbel, 1950). It is worth pointing out that such a mechanism has, in a clear sense, less discrimination than a synaptic one (Eccles, 1953, pp. 219–21). Take a given real neurone with a threshold θ. Then if we alter all input synapses so that the EPSP's or IPSP's which they deliver are increased by a constant multiple μ, this is functionally exactly equivalent to keeping the synapses unchanged but altering θ to $\theta\mu^{-1}$. On the other hand, a set of synaptic changes cannot necessarily be mimicked by threshold changes.

If either or both of these alterations occur the problem then arises as to how they are triggered and how maintained. We shall discuss the latter question later and first talk briefly about the triggering. The problem is that the conventional description of the basis of the action potential and the transmission across synapses, as given in Chapter 2, contains no mention of long-lasting changes. And yet somehow the firing activity in the brain, which forms the primary representation of the material to be remembered, has to initiate some process of change which can probably endure for a lifetime. Short-term processes in memory, that is up to seconds or minutes after an event, could conceivably arise through maintained activity of the kind discussed in relation to Renshaw cells in Section 3.2.2, if ε^{-1} were of the order of seconds or minutes for some cortical cells (Griffith, 1967a; Horridge, 1968). It is unknown if this is true. However, it would seem improbable that long-term memory could be maintained in such a way, i.e. through persistence of transmitter for years.

How the initiation actually occurs is still entirely an open question. One general possibility is that one of the various chemical processes which accompany the action potential or postsynaptic potentials, such as the formation and destruction of transmitter, the transport of sodium and potassium ions, the use of the biochemical energy source adenosine triphosphate, etc. is coupled to some other chemical reaction which controls the production of some compound which itself controls the growth or atrophy of synaptic knobs.

If memory resides in a change of synaptic connection then it would seem quite likely that the initial step leading to a synaptic change should take place at the synapse in question. A rather natural way in which this could happen would be if some chemical "messenger" compound passed across the synapse from one cell to the other whenever both cells fired more or less simultaneously. From the functional point of view this is typically the occasion when the arrival of a presynaptic impulse at that synapse has caused the postsynaptic potential to rise above its threshold to firing. Mechanistically, it could occur if each cellular membrane at the synapse were normally impermeable to the messenger but became permeable to it at the same time as the development of the action potential in its cell. This would not seem an unreasonable possibility in view of the general tendency for superthreshold potential changes to be associated with alteration of membrane permeability. I have called it a "two gates" hypothesis and have discussed it in more detail elsewhere (Griffith, 1966a, 1968c, 1970).

In either case we would next wish to know what sequence of biochemical reactions underlies the later stages of the registering of the long-term memory trace. Here we should naturally frame our questions and speculations in terms of what is known of the biochemistry of cells, especially of nerve cells. Particularly, in view of the continual turnover (destruction and resynthesis) of most cellular constituents, we can ask what in a cell can be sufficiently permanent for the purpose? One possible answer to this is to note that as nerve cells do not normally divide, their DNA should last for a lifetime and chemical alterations of it, perhaps leading to altered rates of protein synthesis, could serve as a stable memory store (Griffith and Mahler, 1969). Another is to note that energy-consuming non-equilibrium systems have, in general, the potentiality of existing in more than one stable steady state. The possible relevance of this fact for biology has been remarked before (Lotka, 1925; Denbigh, Hicks and Page, 1948). We shall now consider one kind of way in which such an idea could be implemented biochemically in relation to memory, starting with a very brief introduction to the biochemistry underlying it. For further biochemical background, the reader may consult Watson (1965), Ingram (1966), Mahler and Cordes (1968).

One hypothetical type of system which can have more than one stable steady state was discussed by Monod and Jacob (1961) in connection with the problem of cellular differentiation. The genetic material, or genes, in a cell consists essentially of DNA, which is a linear polymer built out of four subunits, namely the bases Adenine (A), Guanine (G), Cytosine (C) and Thymine (T), linked in a chain through phosphate (P) and deoxyribose (R) bridges in a manner which may be represented schematically as

```
--- R P R P R P R P R P R P ---
    |   |   |   |   |   |
    G   A   A   C   T   A
```

The genetic information contained in the cell is, at least almost entirely, stored in coded form as the actual sequence of bases. This information tells the cell in what order amino acids should be strung together in proteins. However, it is not used directly, but is first transcribed in sections onto an intermediate polymer called messenger RNA (often abbreviated to mRNA). This is similar to DNA except that the sugar residue R is replaced with a slightly different one, namely ribose (R′) and the four bases are now A, G, C but with Uracil (U) in place of Thymine.

The transcription proceeds through a specific chemical pairing (called base pairing) between G and C (or C and G) and between A and U and T and A as shown schematically as:

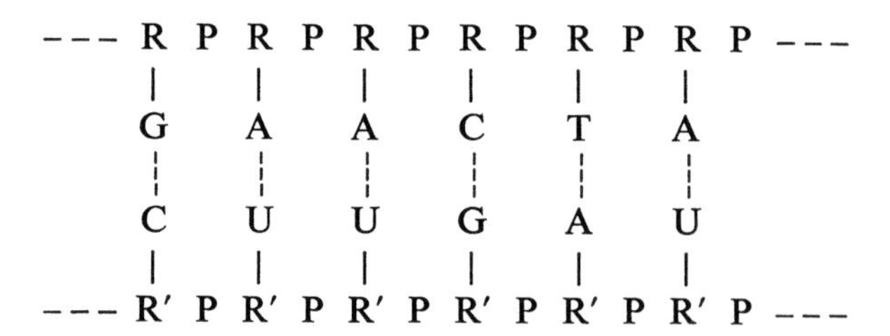

This mRNA then goes as the "machine-controlling instruction tape" to a special protein synthesizing system in the cell where it is "translated" into the amino acid sequences of the proteins which are produced there. Each amino acid gets coded for by a triplet of bases, and this triplet coding scheme is called the genetic code. In the example given above CUU is a triplet for the amino acid leucine, and GAU for aspartic acid. There are, of course, 64 possible triplets, but they are actually coding for only 20 different amino acids (and three instructions controlling the size of the protein). As a consequence several different triplets may correspond to one amino acid. Evidently this presumably implies a loss of information at the translation stage (essentially from $\log 64 = 6$ bits per triplet to $\log 20 = 4.32$ per amino acid). For a mammal, the genetic information content calculated thus is of the order of 10^{10} bits, although exactly how much of this is "used"

in specifying the structure of the animal is a little uncertain (see Brown and Dawid, 1968).

It is probable that the rate of protein synthesis can be controlled in a number of ways. One is that the gene (the relevant section of DNA) may combine with control molecules, probably protein, and be activated or deactivated by them. Those which make the gene more active in mRNA production are called inducers and those which make it less so are repressors.

We are now able to see how there can be alternative steady states by considering two hypothetical situations. The first may be written

$$G \rightarrow M \rightarrow E, \tag{45}$$

where G is the DNA, M the mRNA and E is the protein. Suppose now that E induces G in such a way that when E is absent, G produces no M. Then the system has an "off" state in which there is no M or E and none can be produced because of the absence of the inducer E, and an "on" state in which E indirectly catalyses its own production through inducing G to be transcribed into M. It is not obvious, of course, that the system will really work in the way indicated by this qualitative argument but it is in fact possible for it to do so (see Section 7.2.1 and the end of Section 7.3). Another example of a similar kind is the biochemical flip-flop

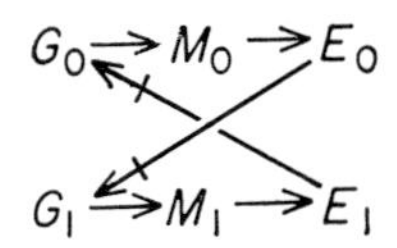

involving interactions between two genes. Here E_0 is hypothesized to repress G_1 and E_1 to repress G_0. Thus when G_0 is "on", the resulting presence of E_0 holds G_1 "off" and vice-versa, i.e. when G_1 is "on", G_0 is "off" (for a mathematical discussion see Section 7.2.2).

A fairly detailed discussion of the way in which bistable or multistable systems of this kind might form a basis for long-term memory has been presented elsewhere (Griffith, 1967a, 1968c) under the name of a switching theory of memory because, in it, the process of recording involves a switching between the possible steady states. We shall not, however, attempt here to discuss this further except for the mathematics given in the next chapter nor shall we try to survey the enormous field of literature relating to the problem of memory. However, it should be apparent from the brief discussion of capacity in the next section that the traditional idea that long-term memory may be expressed in changes of synaptic connection is by no means implausible, although not of course demonstrated experimentally, and that one is not therefore compelled to look for exotic memory mechanisms, as has sometimes been believed. There have been numerous recent papers, reviews and books about the problem of memory and I give here some references to further reading. See Burns (1958), Russell (1959), Hebb (1961), Deutsch (1962), Young (1964, 1966), Richter (1966), Roberts (1966), Dixon (1967), John (1967), Rosenblatt (1967), Bogoch (1968, 1969), Horridge (1968), Glassman (1969), Weisskrantz (1970).

6.5.2. A Simple Formula for Information Capacity

Suppose that the complete contents of a memory store can be specified by giving the values λ_i of n parameters, $i = 1, 2, \ldots, n$, each of which independently could have had one of x possible values. Then the number of different states for the whole store is x^n and its information capacity is $C = n \log_2 x$. We shall be especially interested in the cases in which the parameters are either measures of synaptic strength, in which case λ_i might be the number of synaptic knobs from one cell onto another, or of threshold. In either of these, however, λ_i might either have only a finite number of different possible values, corresponding to a finite number of different control states as discussed in relation to equations (45) and (46), or be effectively continuously variable when we would have the same sort of problem as we had with respect to length of interspike interval in Section 6.3. In the latter case one might try to argue that λ_i had only a limited number x of significantly different ranges. Assuming this is reasonable, then because x occurs as a logarithm, the considerable uncertainty in its value has a relatively small effect upon C and it would be natural to expect $1 \leqslant \log x \leqslant 10$.

With a threshold theory of memory, n is the number of participating nerve cells, so C should be of the order of 10^{10}. In a theory based on change of synaptic connection, n is the number of pairs of cells (a, b) such that modifiable synapses from a to b can normally occur. This could be close to the total number of synapses, or could even be greater, or it might be less by a factor representing the average extent of multiple synapsing between a pair of cells, which could conceivably be at least as great as 60 (Marin-Padilla, 1968). If n is taken equal to the estimated total number of synapses (Section 2.2.6) we then get C of the order of 10^{13}–10^{14}. These figures must be regarded, of course, as extremely crude but at least provide adequate capacity to cope with the final estimates given in Section 6.4.2.

The use of information theory in calculating a memory capacity for the brain is sometimes criticized on the grounds that it is unsatisfactory to apply it to a system about whose internal structure one is so ignorant. We have given a partial defense against this criticism in Section 6.1, although there is some truth in it. However, it does seem to be about the only route through which one can make any quantitative discussion about memory capacity at all.

6.5.3. Distributed Memory

There is a very important and interesting matter that we shall only mention briefly. That is that it seems likely that memory is distributed, that is, that each memory has its trace spread over a large part of the brain. This was indicated originally by the classical experiments of Lashley on rats in which they were trained and then regions of the cortex extirpated. The loss of memory was then roughly proportional to the area extirpated (Lashley, 1929), although the original conclusions have since been modified somewhat in detail. This is to be contrasted with the situation which would hold if each memory were stored in a special localized place, for then extirpation would generally either completely annihilate a given memory or leave it entirely undisturbed. A distributed memory often alarms the biologist, because it implies that memories are all on top of each other, higgledy-piggledy, as if someone were to draw a lot of pictures on top of each other on the same piece of paper. It should not upset the physicist, once he realizes that something akin to a Fourier analysis might easily distinguish the traces. There is no *a priori* reason for supposing the brain would find a Fourier analysis more difficult than a search along a set of localized traces for one having some particular characteristic.

Recently the interesting analogy between this situation, which probably obtains in the brain, and the physical technique of holography, which has many similar features, has attracted much attention (Pribram, 1966; see also Longuet-Higgins, 1968), probably to a great extent because of the difficulty that biologists have had in understanding the possibility of a distributed memory. However, analogs of this kind do also have the advantage of being mathematically fairly tractable and so even if they do not correspond in detail to the neural situation may nevertheless give a helpful insight through yielding definite formulae (see Willshaw, Buneman and Longuet-Higgins, 1969, and also Good, 1966).

CHAPTER 7

Non-linear Equations

7.1. General Introduction

A great number of the equations of motion that we meet in physics are linear or, like the equation of motion

$$a\ddot{\theta} = g \sin \theta$$

for a simple pendulum with length a, may be regarded as approximated by linear equations ($\sin \theta \approx \theta$ for θ small, '.' $= d/dt$). Of course there are notable exceptions, such as Einstein's equations of general relativity theory or, more relevant to our present case, van der Pol's equation

$$\ddot{x} + \mu(x^2 - 1)\dot{x} + x = 0$$

used to describe self-sustaining oscillations in certain electronic multi-vibrator circuits, which are essentially non-linear in the sense that physically important qualitative behavior disappears when the non-linearity is removed.

We shall confine our attention here to non-linear equations, i.e. equations in which the dependent variables (these are $\ddot{\theta}$, θ and so forth) occur non-linearly (as $\sin \theta$ or $x^2\dot{x}$ above), which are also autonomous. The latter means that, like the two equations shown above, they do not contain the independent variable (t) explicitly. This will enable us to discuss mathematically some of the equations involved in the switching theory of memory mentioned at the end of Section 6.5.1. It will also enable us to bring out a number of general features of such equations. The reader who wishes to pursue the subject further can refer to Stoker (1950), Minorsky (1958) or Andronov, Vitt and Khaikin (1966) among others.

Non-linear equations have two features especially of interest to us. One is that they may have oscillatory solutions whose amplitude as well as period are determined by the intrinsic properties of the equations. This is not possible with linear equations because, for them, if $x(t)$ is a solution so also is $cx(t)$ for any constant c. Thus the amplitude depends on the initial conditions. A simple example of a non-linear equation which has just one oscillatory solution, towards which all other solutions tend, is given by Minorsky (1958). He takes the pair of equations

$$\dot{r} = 1 - r^2, \qquad \dot{\theta} = 1$$

which can be converted to a single more complicated equation for x by writing $x = r(\cos\theta - \sin\theta)$, $\dot{x} = r(\cos\theta + \sin\theta)$ and eliminating $\dot{r}$, θ and r in terms of x, $\dot{x}$, $\ddot{x}$. These equations have the solution

$$r = \tanh(t+\alpha), \qquad \theta = t+\beta,$$

with α, β constants. As $t \to \infty$, $r \to 1$ irrespective of the initial conditions. Thus all motions tend to the oscillatory solution $x = \sqrt{2}\cos(t+\beta+\frac{1}{4}\pi)$ which is unique except for the phase $\beta+\frac{1}{4}\pi$. Such a motion is therefore stable with respect to any perturbation of it. Such solutions of non-linear equations, stable at least with respect to small perturbations, often occur and are called stable limit cycles. A system which possesses one stable limit cycle has an intrinsic period and amplitude of oscillation (which need not be sinusoidal). Evidently there is a potential application here to the rhythms which show up in the electroencephalogram (see Grey Walter, 1953; Wiener, 1961; Anderson and Andersson, 1968). However, there is still some argument about whether the dominant human rhythm, the α-rhythm with frequency about 10/second and measured amplitude of the order of 50 microvolts, arises mainly from a summation of individual action potentials, or possibly from slower potentials due to glial cells or even to the relayed potential changes due to tremor in the eye muscles. However, it is worth remarking that if it is an intrinsic oscillation of neural action potential activity and nervous conduction in a region of the brain, then the frequency is not unreasonable. For its period is about 100 msec which is the time taken for a signal to travel and return over 5 cm at 1 meter/second.

A general autonomous linear equation

$$\frac{d^n x}{dt^n} + p_{n-1}\frac{d^{n-1}x}{dt^{n-1}} + \ldots + p_1 x + p_0 = 0$$

has, if $p_1 \neq 0$, one unique constant solution, namely $x = -p_0 p_1^{-1}$. However, an autonomous non-linear equation may have more than one. In fact, it may even have two or more which are stable to small perburbations (see Section 7.2). Put more physically, a system described by such an equation may be able to persist in any one of two or more distinct stable steady-state conditions. This in fact occurs for the equations describing electronic bistable flip-flop circuits and, as we shall see shortly, for those describing the biochemical control situations discussed in Section 6.5.1. It was probably first realized clearly by Lotka (1925) and Volterra (1931) that such equations might underlie many situations in biology where a system has two or more potentially stable conditions.

7.2. Illustrative Examples

7.2.1. Positive Feedback to One Gene

We now show that it is possible for the system in which a gene is directly induced by the protein E for which it codes (cf. equation (6.45)) to have two stable steady states. This gives a convenient introduction to the phase plane techniques which are very useful generally in the theory. Let the inducer E combine with the gene according to the equation

$$G+mE = GE_m, \tag{1}$$

where G is inactive to produce messenger and GE_m is the active form. It follows from equation (1) that the fraction of the time that G is active is given by

$$p = \frac{KE^m}{1+KE^m}, \tag{2}$$

where E is also the concentration of the protein coded for by G and K is the equilibrium constant for reaction (1). Actually equation (2) assumes that the concentration of G (perhaps only 2 molecules/cell) is small compared with E and should be corrected near $E = 0$. This is unlikely to affect our conclusions significantly and we ignore it here (see Griffith, 1968a).

Then if M is the concentration of the corresponding mRNA, it seems reasonable to write the equations

$$\begin{aligned} \dot{M} &= \frac{aKE^m}{1+KE^m} - bM, \\ \dot{E} &= cM - dE, \end{aligned} \tag{3}$$

which assume that the rate of messenger production is proportional to p and of protein production is proportional to messenger concentration. They also assume that messenger and protein decay by first order processes. We can now simplify these equations by changing units. Write $M \to \mu M$, $E \to \varepsilon E$, $t \to \tau t$, choosing $K\varepsilon^m = c\mu\tau\varepsilon^{-1} = \tau a\mu^{-1} = 1$. Also define $\alpha = b\tau$, $\beta = d\tau$, whereupon

$$\begin{aligned} \dot{M} &= \frac{E^m}{1+E^m} - \alpha M, \\ \dot{E} &= M - \beta E, \end{aligned} \tag{4}$$

with $\alpha > 0$, $\beta > 0$. These could easily be turned into a single second-order equation for E but it is more convenient for most purposes to keep them as they are.

The steady state solutions occur when $\dot{M} = \dot{E} = 0$, i.e. when

$$E^m = \alpha M(1+E^m) = \alpha\beta E(1+E^m). \tag{5}$$

Hence $E = M = 0$ is one such solution and any others must satisfy

$$E^{m-1} = \alpha\beta(1+E^m). \tag{6}$$

When $m = 1$ this means that $E = (\alpha\beta)^{-1}-1$. Because of the physical significance of M and E, we are obviously then only interested in the case $E \geqslant 0$ which occurs if $\alpha\beta \leqslant 1$.

All cases $m \geqslant 2$ behave in a qualitatively similar way (Griffith, 1968a) but for ease of exposition we shall only discuss $m = 2$ here because we can then solve equation (6) explicitly to give

$$E = \frac{1 \pm (1-4\alpha^2\beta^2)^{\frac{1}{2}}}{2\alpha\beta}. \tag{7}$$

Thus there are two real solutions $E > 0$ if $2\alpha\beta < 1$, one if $2\alpha\beta = 1$ and none if $2\alpha\beta > 1$. This preliminary investigation of the steady state solutions has yielded a number of cases, which we treat in turn.

Case 1. $m = 1$, $\alpha\beta > 1$.

There is only one steady state solution, namely $E = M = 0$. All other solutions tend to this one as $t \to \infty$ as we readily see by considering the properties of the function

$$L = M + \alpha E. \tag{8}$$

Using equations (4) we have

$$\begin{aligned} \dot{L} &= E/(1+E) - \alpha M + \alpha(M - \beta E) \\ &= E(1 - \alpha\beta - \alpha\beta E)/(1+E), \end{aligned}$$

which, because $\alpha\beta > 1$, is negative for all $E > 0$. If initially $E = 0$, $M > 0$, then $\dot{E} = \mathrm{M} > 0$ and so E immediately becomes positive and $\dot{L}$ negative. Therefore L decreases continuously. The only point at which L can remain zero is at $E = M = 0$, where $L = 0$, so we have shown that $L \to 0$. However, because $\alpha > 0$ and $M \geqslant 0$, $E \geqslant 0$ we can deduce from this and the definition of L that E and M separately tend to zero. This establishes the result we wish to prove, which is that $E = M = 0$ is a unique stable steady state. It is clear also that this remains true when $\alpha\beta = 1$.

Two general points in the theory of non-linear equations may be illustrated by means of this case. One is the phase-plane representation. For any pair of values of M and E equation (4) gives $\dot{M}$ and $\dot{E}$ and hence the direction in the (M, E) plane in which the values of M and E move. The (M, E) plane is called the phase plane, in analogy with the phase space of classical statistical mechanics, and a point in it represents a particular state for the system, namely it determines the two concentrations M and E. We may then plot in the plane the trajectories along which the representative points move. This is shown schematically in Fig. 7.1(a) and shows very graphically that

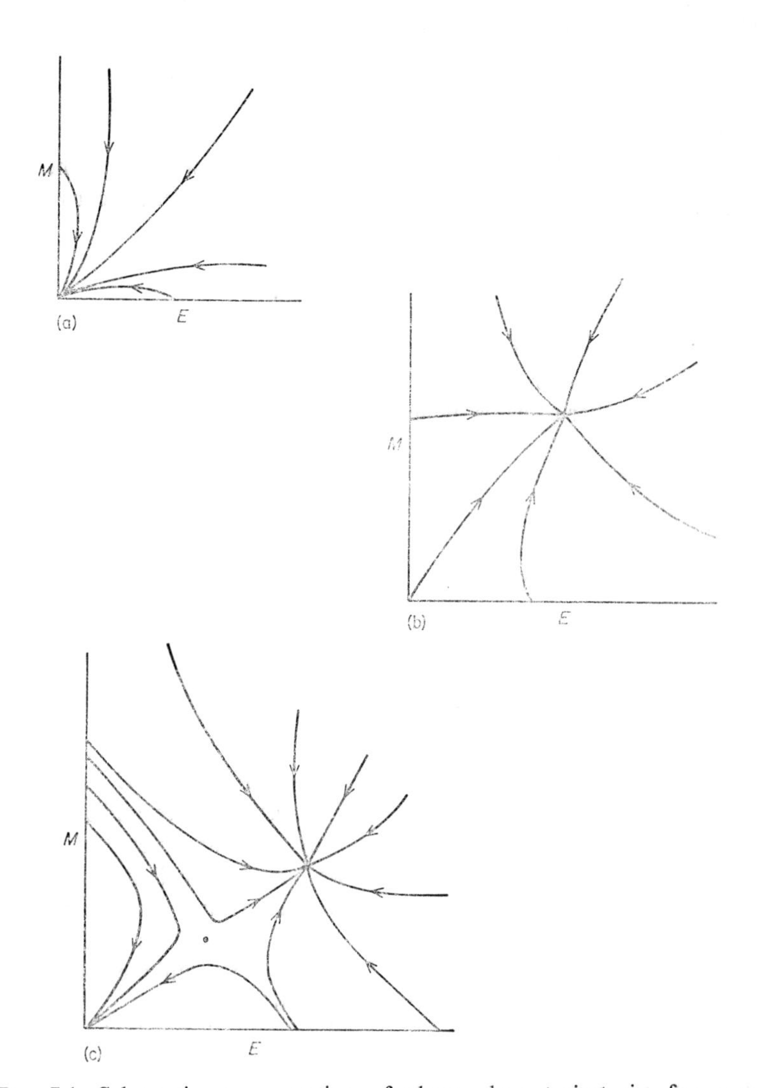

FIG. 7.1. Schematic representation of phase plane trajectories for equations governing positive feedback to one gene considered in Section 7.2.1.
(a), $m = 1$, $\alpha\beta > 1$ or $m = 2$, $2\alpha\beta > 1$; (b), $m = 1$, $\alpha\beta < 1$; (c), $m = 2$, $2\alpha\beta < 1$.

the point $M = E = 0$ is a stable state towards which all others tend. The analysis of non-linear equations through their steady state solutions and phase-plane trajectories furnishes a very powerful technique in the theory.

The other general point emerges from the use of the function L. This is a very special case of Lyapunov's method (Hahn, 1963). In it, a function is found which can be shown, by using the equations of motion, to tend to zero and which has a functional form such that when it tends to zero one can deduce that all the dependent variables of the original equations tend to one of their steady state sets of values. There are two dependent variables in equation (4). If there are more, the phase plane type of method loses most of its power, but Lyapunov's method is often still useful.

Case 2. $m = 2$, $2\alpha\beta > 1$.

This comes out in a very similar way. Here we have

$$\dot{L} = E^2/(1+E^2)-\alpha\beta E$$
$$= -\frac{E}{\alpha\beta(1+E^2)}[(\alpha\beta-\tfrac{1}{2}E)^2+\tfrac{1}{4}(4\alpha^2\beta^2-1)E^2], \qquad (9)$$

which again has the required properties to show that $M = E = 0$, which is the only steady state, is stable. When $2\alpha\beta = 1$ there is, in addition, a metastable steady state at $M = 1/(2\alpha)$, $E = 1/(2\alpha\beta)$.

Case 3. $m = 1$, $\alpha\beta < 1$.

There are two steady states, namely $M = E = 0$ and $M = \alpha^{-1}-\beta$, $E = (\alpha\beta)^{-1}-1$. We now examine the behavior in the neighborhood of them. First, near the origin we have

$$\dot{M} \doteqdot E-\alpha M,$$
$$\dot{E} = M-\beta E.$$

For a reason which will be apparent in a moment, we prefer to investigate the more general equations

$$\dot{M} = \gamma E-\alpha M$$
$$\dot{E} = \eta M-\beta E, \qquad (10)$$

with $\alpha > 0$, $\beta > 0$ and where γ and η are arbitrary real constants which we subsequently set equal to unity. Now differentiate the equation for $\dot{E}$:

$$\ddot{E} = \eta\dot{M}-\beta\dot{E} = (\beta^2+\eta\gamma)E-(\alpha+\beta)\eta M$$

and eliminate M to give

$$\ddot{E}+(\alpha+\beta)\dot{E}+(\alpha\beta-\eta\gamma)E = 0. \qquad (11)$$

This is a linear equation with constant coefficients which we solve by setting $E = e^{\lambda t}$, whence

$$\lambda = -\tfrac{1}{2}(\alpha+\beta) \pm \tfrac{1}{2}[(\alpha+\beta)^2 + 4(\eta\gamma - \alpha\beta)]^{\frac{1}{2}}$$
$$= -\tfrac{1}{2}(\alpha+\beta) \pm \tfrac{1}{2}[(\alpha-\beta)^2 + 4\eta\gamma]^{\frac{1}{2}}. \qquad (12)$$

The point $M = E = 0$ is stable if and only if the real parts of both values of λ are negative which occurs if and only if $\eta\gamma < \alpha\beta$. In our case $\eta = \gamma = 1$ and $\alpha\beta < 1$ so this condition is not satisfied.

However, the other point is stable and we can use the same analysis to help us to show this. Write $E_0 = (\alpha\beta)^{-1} - 1$, $M_0 = \beta E_0$. Then near this point we have

$$M = M_0 + x$$
$$E = E_0 + y,$$

where x and y are both small. Hence, remembering that $(1+E_0)^{-1} = \alpha\beta$,

$$\dot{x} = \dot{M} = (E_0+y)/(1+E_0+y) - \alpha M_0 - \alpha x$$
$$= (E_0+y)(1+E_0)^{-1}(1+y/(1+E_0))^{-1} - \alpha M_0 - \alpha x$$
$$\doteqdot \alpha\beta(E_0+y)(1-\alpha\beta y) - \alpha\beta E_0 - \alpha x$$
$$\doteqdot \alpha^2\beta^2 y - \alpha x$$
$$\dot{y} = x - \beta y.$$

We can now use our previous analysis with $\gamma = \alpha^2\beta^2$, $\eta = 1$ to say that the point is stable if and only $\alpha^2\beta^2 < \alpha\beta$, which it is. All initial concentrations tend towards this stable stationary pair (M_0, E_0), see Fig. 7.1(b), although we have not proved this rigorously here.

Case 4. $m = 2$, $2\alpha\beta < 1$.

This is treated in a manner similar to the last. Near $M = E = 0$, we have now

$$\dot{M} \doteqdot -\alpha M$$
$$\dot{E} = M - \beta E$$

and so the condition for stability $(0 \times 1 < \alpha\beta)$ is always satisfied. The other two steady state values for E may be written (cf. equation (7)) as

$$0 < E_1 < 1/(2\alpha\beta) < E_2.$$

Again, set $M = \beta E_i + x$, $E = E_i + y$, where $i = 1, 2$. We easily find that

$$\dot{x} \doteqdot 2\alpha\beta(1-\alpha\beta E_i)y - \alpha x,$$
$$\dot{y} = x - \beta y.$$

So the condition for the point to be stable reads

$$2\alpha\beta(1-\alpha\beta)E_i < \alpha\beta,$$

which can be rearranged to $E_i > 1/(2\alpha\beta)$. Hence $(\beta E_1, E_1)$ is unstable and $(\beta E_2, E_2)$ is stable. This is illustrated in Fig. 7.1(c).

7.2.2. A Biochemical Flip-flop

We now give an analysis of a simplified version of the flip-flop control system discussed in Section 6.5.1. In place of equation (1) of the last subsection, we now have

$$G_0 + E_0 = G_0 E_0,$$
$$G_0 + E_1 = G_0 E_1, \tag{13}$$

where only $G_0 E_0$ produces the corresponding messenger M_0. There are similar equations for G_1. There is now competition between E_0 and E_1 for the genes G_0 and G_1 which results in equation (2) being replaced with

$$p_0 = \frac{K_0 E_0}{1 + K_0 E_0 + l_0 E_1} \tag{14}$$

for the fraction of the time that G_0 is combined with E_0. K_0 and l_0 are respectively the equilibrium constants of the two equations (13). As a consequence we can write

$$\dot{M}_0 = \frac{a_0 K_0 E_0}{1 + K_0 E_0 + l_0 E_1} - b_0 M_0,$$
$$\dot{E}_0 = c_0 M_0 - d_0 E_0, \tag{15}$$

with similar equations for M_1 and E_1.

We now have a four-variable system which we shall reduce to a two-variable one by only considering the case in which c_0, d_0 and c_1, d_1 are very large so that E_0 and E_1 always adjust rapidly to the prevailing values of M_0 and M_1. Then we can deduce that $E_0 = c_0 d_0^{-1} M_0$, $E_1 = c_1 d_1^{-1} M_1$. There are then two equations for M_0 and M_1 which, by an alteration of units can be written as

$$\dot{M}_0 = \frac{\gamma_0 M_0}{1 + M_0 + l_0 M_1} - \alpha_0 M_0,$$
$$\dot{M}_1 = \frac{\gamma_1 M_1}{1 + M_1 + l_1 M_0} - \alpha_1 M_1, \tag{16}$$

where all the constants γ_0, α_0, l_0, etc., are positive.

Equations (16) have four steady state solutions. It is convenient to set $\xi_0 = \gamma_0/\alpha_0$, $\xi_1 = \gamma_1/\alpha_1$, whereupon they become

1. $M_0 = M_1 = 0$,
2. $M_0 = 0$, $M_1 = \xi_1 - 1$,
3. $M_1 = 0$, $M_0 = \xi_0 - 1$,
4. The solutions of $1 + M_0 + l_0 M_1 = \xi_0$, $1 + M_1 + l_1 M_0 = \xi_1$, m_0 and m_1 say.

Stability of these points is easily investigated.

Solution 1. Near the origin we have

$$\dot{M}_0 \doteqdot (\gamma_0-\alpha_0)M_0,$$
$$\dot{M}_1 \doteqdot (\gamma_1-\alpha_1)M_1,$$

and so the origin is stable if and only if $\alpha_0 > \gamma_0$, $\alpha_1 > \gamma_1$, which can be rearranged into the form $\xi_0 < 1$, $\xi_1 < 1$.

Solution 2. This exists physically if and only if $M_1 \geqslant 0$, i.e. $\xi_1 \geqslant 1$. Then we set $M_1 = \xi_1 - 1 + x_1$, with x_1 small, and readily obtain

$$\dot{M}_0 \doteqdot \left(\frac{\gamma_0}{1+l_0(\xi_1-1)} - \alpha_0\right) M_0,$$
$$\dot{x}_1 = \dot{M}_1 \doteqdot -\alpha_2(1-\xi_1^{-1})(x_1+l_1 M_0).$$

For stability it is clearly necessary that the coefficient of M_0 in the first equation should be negative, i.e. that $1+l_0(\xi_1-1) > \xi_0$, if we assume that $\xi_1 \geqslant 1$. It then follows from the analysis given previously of equation (10) that this condition is also sufficient.

Solution 3. This exists physically for $\xi_0 \geqslant 1$ and is stable for

$$1+l_1(\xi_0-1) > \xi_1.$$

Solution 4. Here we set $M_0 = m_0 + x_0$, $M_1 = m_1 + x_1$ and find that

$$\dot{x}_0 = \dot{M}_0 \doteqdot -\alpha_0 \xi_0^{-1}(x_0 + l_0 x_1),$$
$$\dot{x}_1 = \dot{M}_1 \doteqdot -\alpha_1 \xi_1^{-1}(x_1 + l_1 x_0).$$

This is stable if and only if

$$\alpha_0 \xi_0^{-1} l_0 \cdot \alpha_1 \xi_1^{-1} l_1 < \alpha_0 \xi_0^{-1} \cdot \alpha_1 \xi_1^{-1},$$

i.e. when $l_0 l_1 < 1$. If we assume this is true then as

$$m_0 = (1-l_0 l_1)^{-1}(\xi_0 - l_0 \xi_1 + l_0 - 1),$$
$$m_1 = (1-l_0 l_1)^{-1}(\xi_1 - l_1 \xi_0 + l_1 - 1),$$

it follows that if these are positive, then

$$\xi_0 - 1 > l_0(\xi_1 - 1),$$
$$\xi_1 - 1 > l_1(\xi_0 - 1), \tag{17}$$

whence $(\xi_1 - 1) > l_0 l_1(\xi_1 - 1)$ which, as $l_0 l_1 < 1$, means that $\xi_1 \geqslant 1$. Similarly $\xi_0 \geqslant 1$. Therefore, we have shown that if solution 4 exists and is stable, then solutions 1, 2 and 3 all exist and are unstable. There is, in fact, an exclusion principle operating in relation to stability, for our results also show that if solution 1 is stable then solutions 2, 3 and 4 do not exist. If either solution 2 or 3 is stable, then solutions 1 and 4 (if it exists) are necessarily unstable. The bistable flip-flop situation occurs when both solutions 2 and 3 are stable, which is possible as we can see by putting

$$\xi_0 = \xi_1 = l_0 = l_1 = 2,$$

for example.

7.2.3. Oscillations

Goodwin (1963) has remarked that if we replace the right-hand sides of equations (4) with $(1+E)^{-1}-\alpha$ and $M-\beta$, respectively, the resultant equations have oscillatory solutions. He shows this by combining them as

$$(M-\beta)\dot{M}-(1/(1+E)-\alpha)\dot{E} = 0,$$

which is immediately integrable to give

$$G(M,E) \equiv \tfrac{1}{2}M^2-\beta M-\ln(1+E)+\alpha E = \text{constant}. \qquad (18)$$

Thus the phase-plane motion is along curves of constant G. There is just one steady state solution, namely $M = \beta$, $E = \alpha^{-1}-1$ (we assume $\alpha < 1$) and G is constant along closed curves surrounding this point. Those which are near to it are approximately small ellipses as we see by putting $M = \beta+x$, $E = \alpha^{-1}-1+y$, where x and y are small, because then $\dot{x} \doteqdot -\alpha^2 y$, $\dot{y} = x$, whence $x^2+\alpha^2y^2 = \text{constant}$. This is shown schematically in Fig. 7.2. However, this example is rather artificial in that when the constant in equation (18) is large, the curves cross from the physically allowed region of the plane ($M \geqslant 0$, $E \geqslant 0$) to the unacceptable region ($M < 0$ or $E < 0$). This should be borne in mind when considering the general scheme Goodwin (1963) has founded on this type of equation (see also Griffith, 1968a).

A mathematically similar example was given by Lotka (1920), who considered the following hypothetical autocatalytic reaction scheme:

$$A \xrightarrow{k_1} B \xrightarrow{k_2} C \xrightarrow{k_3},$$

where A is maintained at a constant concentration $\bar{a}$. C catalyses its own production as is apparent in the equations

$$\dot{b} = k_1\bar{a}b-k_2bc,$$
$$\dot{c} = k_2bc-k_3c, \qquad (19)$$

where b and c are the concentrations of B and C respectively. Again there is an integral of the motion which is obtained by dividing one equation by the other, thus

$$\frac{db}{dc} = \frac{b(k_1\bar{a}-k_2c)}{c(k_2b-k_3)}$$

which integrates to

$$L(b,c) \equiv k_2b-k_3\ln b-k_1\bar{a}\ln c+k_2c = \text{constant}. \qquad (20)$$

Figure 7.2 serves schematically for this case too, showing that the concentrations of b and c oscillate about their steady state values which are $k_3k_2^{-1}$ and $k_1\bar{a}k_2^{-1}$ respectively. In this case, however, any curve (20) which starts in the quadrant $b \geqslant 0$, $c \geqslant 0$ stays there, so there is nothing physically objectionable about equations (19). Note, however, that Lotka asserts that the period of the oscillation is the same, independently of the starting values for b and c, and is always $T = 2\pi/(k_1 k_3 \bar{a})$. This is untrue and in fact the period is always greater than T, although it tends to this value as the amplitude of the oscillation tends to zero.

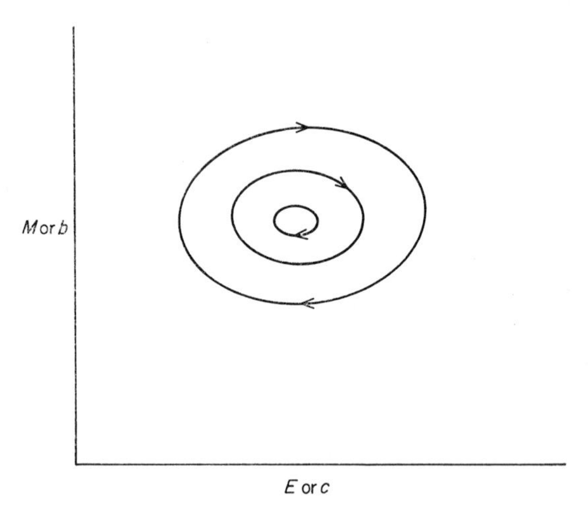

Fig. 7.2. Schematic representation of phase plane trajectories for Goodwin's and Lotka's equations (Section 7.2.3). The behavior near the axes is not the same in the two cases.

Because of the superficial analogy of the flip-flop of Section 7.2.2 to an electronic bistable circuit, one might naturally ask if there exist similar hypothetical biochemical analogies to electronic multivibrator circuits. Specifically one might consider a control situation of the type (→ symbolizes induction and ↛ repression)

$$\begin{array}{c} G_0 \rightarrow M_0 \rightarrow E_0 \\ G_1 \rightarrow M_1 \rightarrow E_1 \end{array}$$

where one protein E_0 induces G_1 while E_1 represses G_0. However, I have found it very difficult to obtain oscillations using physiochemical equations constructed as in Sections 7.2.1 and 7.2.2, i.e.

$$\dot{M}_0 = (1 + aE_1^m)^{-1} - \alpha M_0,$$
$$\dot{E}_0 = M_0 - \beta E_0,$$
$$\dot{M}_1 = E_0^m(1 + bE_0^m)^{-1} - \gamma M_1,$$
$$\dot{E}_1 = M_1 - \varepsilon E_1.$$

This was only investigated by computer simulation but for quite a wide range of values of the constants and no sustained oscillations were found.

The same applied to the control situation

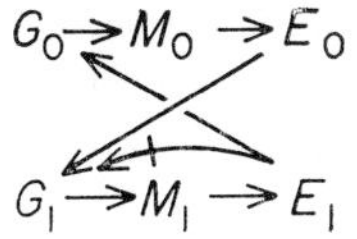

$$G_0 \to M_0 \to E_0$$
$$G_1 \to M_1 \to E_1$$

and only with

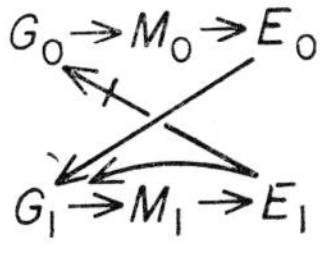

$$G_0 \to M_0 \to E_0$$
$$G_1 \to M_1 \to E_1$$

were any oscillations found. Taking the equations

$$\begin{aligned}
\dot{M}_0 &= (1+E_1^3)^{-1} - \tfrac{1}{4}M_0, \\
\dot{E}_0 &= M_0 - \tfrac{1}{4}E_0, \\
\dot{M}_1 &= E_0^3 E_1^3 (1+E_0^3 E_1^3)^{-1} - \tfrac{1}{4}M_1, \\
\dot{E}_1 &= M_1 - \tfrac{1}{4}E_1,
\end{aligned}$$

a limit cycle was found, and checked for stability, passing through

$$M_0 = 4.00, \qquad E_0 = 1.60, \qquad M_1 = 1.16, \; E_1 = 4.62.$$

The existence of integrals of the motion, as given in equations (18) and (20), occurs naturally in isolated systems with a law of conservation of energy (cf. the simple pendulum). Unless there is some such general reason for them, they will not usually occur with a non-linear system. Therefore we are probably wisest to regard the occurrence of $G(M, E)$ and $L(b, c)$ as being either contrived or accidental. Therefore, also, the infinite set of oscillations having varying amplitude and corresponding to different values of the constant of the motion, G or L, should be regarded as exceptional in the present context.

Oscillations have been found in biochemical systems, although it does not seem quite clear whether these are of the limit cycle type or are very slowly damped oscillations tending towards a steady state (Chance, Estabrook and Ghosh, 1964; Pye and Chance, 1966; Higgins, 1967; Betz, 1968). Furthermore, Horn (1962) found certain nerve cells in the visual cortex of cat which fired rather regularly with a period of the order of 10 sec and suggested that they might have been firing in relation to some underlying biochemical oscillation. Obviously this is speculative, but is perfectly possible.

7.2.4. The Bendixson Criterion

It is often possible to show easily and definitely that there can be no oscillatory solutions to a pair of equations

$$\begin{aligned}
\dot{M} &= P(M, E), \\
\dot{E} &= Q(M, E).
\end{aligned}$$

This is done by hypothesizing the existence of an oscillation, i.e. a closed trajectory l in the (M, E) plane, and applying Stokes's theorem (see Rutherford, 1954, p. 74) to the interior S of this trajectory. We have

$$\int_S \left(\frac{\partial P}{\partial M} + \frac{\partial Q}{\partial E}\right) dS = \int_l (P\, dE - Q\, dM).$$

But $P\, dE = P\dot{E}\, dt = \dot{M}\dot{E}\, dt = Q\dot{M}\, dt = Q\, dM$, so the right-hand side is zero. Hence if we can show that the left-hand side is non-zero we can deduce that our initial hypothesis of the existence of a closed trajectory is false.

Bendixson's criterion is a sufficient condition for this deduction, namely that $\partial P/\partial M$ and $\partial Q/\partial E$ should both be always positive, or both be always negative, over the whole plane. For example, equations (4) cannot yield oscillations because $\partial P/\partial M = -\alpha$, $\partial Q/\partial E = -\beta$.

7.3. The Hurwitz Criteria

We have only discussed the stability of steady states for equations with two dependent variables. However, even if there are more, one can still expand about the steady state values and test for stability. Suppose

$$\dot{x}_i = f_i \quad (x_1, x_2, \ldots, x_n), \tag{21}$$

where $i = 1, 2, \ldots, n$. Then the steady states are the solutions of the set of simultaneous equations

$$f_i(x_1, x_2, \ldots, x_n) = 0, \quad i = 1, 2, \ldots, n.$$

Let one such state be $x_1^0, x_2^0, \ldots, x_n^0$. Then set

$$x_i = x_i^0 + y_i,$$

where y_i is small. Evidently

$$\dot{y}_i = \dot{x}_i \doteqdot \sum_{j=1}^{n} \left(\frac{\partial f_i}{\partial x_j}\right)_0 y_j, \tag{22}$$

where $\partial f_i/\partial x_j$ is evaluated for the steady state values. Equations (22) are linear equations with constant coefficients and so we try $y_i = a_i\, e^{\lambda t}$. Setting $h_{ij} = (\partial f_i/\partial x_j)_0$, we obtain

$$\lambda a_i = \sum_{j=1}^{n} h_{ij} a_j$$

and so λ is one of the eigenvalues of the matrix $H = [h_{ij}]$. In the usual case that there are n different values of λ, we then have n independent non-zero solutions to (22), in terms of which any solution may be expanded, and the point x_i^0 is stable when all the real parts of the n eigenvalues λ are negative.

Therefore we wish to know the conditions under which the real parts of all roots of a characteristic equation

$$F(\lambda) \equiv \lambda^n + p_{n-1}\lambda^{n-1} + \ldots + p_1\lambda + p_0 = 0 \tag{23}$$

are negative. We shall discuss the cases $n = 2$ and 3 in detail, by simple elementary arguments, and refer elsewhere for the general case. First note that all complex roots of equation (23) occur in conjugate pairs. This is because if $F(\lambda) = 0$ we can immediately deduce that $F(\lambda)^* = F(\lambda^*) = 0$. If $\lambda_1, \lambda_2, \ldots, \lambda_n$ are the roots, then

$$F(\lambda) \equiv \prod_{i=1}^{n} (\lambda - \lambda_i). \tag{24}$$

When λ_1, λ_2 are a conjugate pair $-\xi \pm i\eta$ the product

$$(\lambda - \lambda_1)(\lambda - \lambda_2) = \lambda^2 + 2\xi\lambda + \xi^2 + \eta^2,$$

while if $\lambda_3 = -\alpha$ is real, $(\lambda - \lambda_3) = \lambda + \alpha$, which enables us to deduce that a necessary condition for all real parts of the roots of equation (23) to be negative is that all the coefficients p_j $(j = 0, 1, \ldots, n-1)$ must be positive.

When $n = 2$, this condition is also sufficient. For if $p_0 > 0$, $p_1 > 0$, then as

$$\lambda = -\tfrac{1}{2}p_1 \pm \tfrac{1}{2}(p_1^2 - 4p_0)^{\frac{1}{2}},$$

the result follows at once.

For $n = 3$, the conditions $p_0 > 0$, $p_1 > 0$, $p_2 > 0$ are not sufficient. A counterexample is the equation $x^3 + \frac{1}{2}x^2 + \frac{1}{2}x + 1 \equiv (x+1)(x^2 - \frac{1}{2}x + 1)$. The analysis here can be made to depend on the function $D_2 = p_1 p_2 - p_0$, necessary and sufficient conditions being that $p_0 > 0$, $p_1 > 0$ and $D_2 > 0$, as we now see. The function

$$F(\lambda) \equiv \lambda^3 + p_2\lambda^2 + p_1\lambda + p_0$$

satisfies $F(-\infty) = -\infty$, $F(+\infty) = +\infty$ so it always has one real root $F(\alpha) = 0$. Therefore it can be factorized as

$$F(\lambda) \equiv (\lambda + \alpha)(\lambda^2 + a\lambda + b)$$

with a and b real. We have already dealt with the case $n = 2$ which shows that the necessary and sufficient condition on $F(\lambda)$ is that $\alpha > 0$, $a > 0$, $b > 0$. But

$$p_0 = \alpha b, \qquad p_1 = \alpha a + b, \qquad p_2 = a + \alpha, \qquad D_2 = a(\alpha^2 + \alpha a + b),$$

and so this implies that all of p_0, p_1, p_2 and D_2 are positive. Conversely, assume that $p_0 > 0$, $p_1 > 0$, $D_2 > 0$. Then $p_0 > 0$ means that either $\alpha > 0$, $b > 0$ or $\alpha < 0$, $b < 0$. If the latter is true, then $p_1 > 0$ implies that $a < 0$ also. But then $D_2 = a(\alpha^2 + p_1) < 0$ which is a contradiction. Hence $\alpha > 0$, $b > 0$. $D_2 > 0$ now implies that $a > 0$ which proves the result. Alternative necessary and sufficient conditions are $p_0 > 0$, $p_2 > 0$, $D_2 > 0$, which is easily proved in the same way. However, $p_1 > 0$, $p_2 > 0$, $D_2 > 0$ is not sufficient.

We have just proved two special cases of Hurwitz's criteria (see Uspensky, 1948; Porter, 1967). For general n, these criteria give necessary and sufficient conditions that all real parts of the λ_i should be negative. They are that $p_0 > 0$ and that each of the $n-1$ determinants

$$D_1 = p_1, \quad D_2 = \begin{vmatrix} p_1 & p_0 \\ p_3 & p_2 \end{vmatrix},$$

$$D_3 = \begin{vmatrix} p_1 & p_0 & 0 \\ p_3 & p_2 & p_1 \\ p_5 & p_4 & p_3 \end{vmatrix}, \quad \ldots, \quad D_{n-1} = \begin{vmatrix} p_1 & p_0 & \cdots & 0 \\ p_3 & p_2 & \cdots & 0 \\ \cdot & \cdot & \cdot & \cdot \\ p_{2n-3} & p_{2n-4} & \cdots & p_{n-1} \end{vmatrix}$$

should be positive. In the D_j, we set $p_n = 1$ and all p_l having $l > n$ equal to zero.

As an illustration of the use of Hurwitz's criteria we consider the generalization

$$\begin{aligned} \dot{M} &= \frac{P^m}{1+P^m} - \alpha M, \\ \dot{E} &= M - \beta E, \\ \dot{P} &= E - \gamma P, \end{aligned} \tag{25}$$

of equations (4). The steady state solutions satisfy an equation almost identical with (5), namely

$$P^m = \alpha\beta\gamma P(1+P^m).$$

All steady states have stability properties exactly matching those for the two variable case. We shall just discuss the point $M = E = P = 0$, but the others can be dealt with almost equally easily. Near this point, equations (25) reduce to

$$\begin{aligned} \dot{M} &\doteqdot P\delta_{m1} - \alpha M, \\ \dot{E} &= M - \beta E, \\ \dot{P} &= E - \gamma P, \end{aligned}$$

where $\delta_{m1} = 1$ if $m = 1$, and is zero otherwise. The matrix H is thus

$$H = \begin{bmatrix} -\alpha & 0 & \delta_{m1} \\ 1 & -\beta & 0 \\ 0 & 1 & -\gamma \end{bmatrix}$$

whose characteristic equation is

$$\lambda^3 + \lambda^2 \sum \alpha + \lambda \sum \alpha\beta + \alpha\beta\gamma - \delta_{m1} = 0. \tag{26}$$

Hence

$$p_2 = \sum \alpha > 0, \qquad p_1 = \sum \alpha\beta > 0, \qquad p_0 = \alpha\beta\gamma - \delta_{m1}$$

and

$$\begin{aligned} D_2 = p_1 p_2 - p_0 &= (\textstyle\sum \alpha)(\sum \alpha\beta) - \alpha\beta\gamma + \delta_{m1} \\ &= \textstyle\sum \alpha^2\beta + 2\alpha\beta\gamma + \delta_{m1} > 0. \end{aligned}$$

So the point is stable for $m > 1$. For $m = 1$, it is stable if $\alpha\beta\gamma > 1$, but not if $\alpha\beta\gamma < 1$. Note that when $m > 1$ we do not actually need the Hurwitz criteria to discuss equation (26) because then $\lambda = -\alpha, -\beta, -\gamma$ are the actual solutions. Also note that $L = M+\alpha E+\alpha\beta P$ serves as a possible Lyapunov function for some values of the parameters in equation (25) and this gives us the easiest way to show that $M = E = P = 0$ is also stable for $\alpha\beta\gamma = 1$ when $m = 1$ (for further discussion see Griffith, 1968a).

CHAPTER 8

General Questions about the Brain

8.1. Some Experimental Matters

In this section I shall draw attention to a number of observed general features especially of the mammalian brain which seem to me to be worth considering as possible pointers to the principles underlying its operation. The interested reader will naturally wish to supplement my very brief summaries with reference to the original literature in order to form his own opinion of whether I have correctly emphasized the most salient features. He will also probably realize that I have recently published elsewhere a speculative theory of brain organization (see Griffith, 1967a, b, c; 1968b). I do not wish to repeat it here but merely to point out that all of the matters raised in this section (and also Section 6.5.3) were specifically incorporated into it.

8.1.1. Relative Insensitivity to Damage

One of the most remarkable features of the brain is the relatively small effect on its capabilities of even quite a large destruction of neurones. The magnitude of the effect depends very much on the position of the destruction and is greatest in sub-cortical regions, such as the thalamus, or in regions of the cortex which are specialized as primary sensory receiving areas, as for example the visual cortex, or primary output areas, especially the motor cortex through which most commands to the musculature pass. But quite large parts of the rest of the cortex can be removed with comparatively little observable effect on the capabilities of the animal or of man (see Russell, 1959). Perhaps the most famous example of this is the operation of prefrontal leucotomy in which the most forward part of the cortex, on both sides, is surgically separated from the rest of the brain. There are, of course, known to be changes in personality as a result of this operation; what should impress us, however, is that the brain works at all after it.

It has been estimated that something like 10^5 neurones in the cortex die per day (LeGros Clark, 1963, p. 160). It is probably generally believed that this occurs comparatively at random, thus furnishing further evidence of insensitivity to damage, but the possibility that the cells which die are carefully selected by the brain, perhaps as part of a memory system based on the resulting reduction in connections, cannot be entirely ruled out.

Another piece of evidence comes from the work of Sperry (1947) and Sperry and Miner (1955) who took monkeys and criss-crossed their cerebral cortices with large numbers of cuts at right angles to the surface, without producing any significant lasting effect on their memory or behavior after they had recovered from the operations.

In a somewhat similar, although more startling, vein, are the split-brain preparations of Sperry (1961). In these, the corpus callosum and other structures connecting the two halves of the brain are completely severed to a great depth by an incision along the plane of symmetry of the brain. Remarkably, each of the two half-brains functions by itself. In the case of man, one half, called the dominant half and usually found on the left, appears to be essentially as capable of thought and function as the whole brain was before, whilst the other, non-dominant, half does function but at a much lower level of capability. Other mammals appear to be less specialized in this respect and both halves separately each work about as well as each other and as the original undivided brain. Where one had one integrated functioning system before one has, by the stroke of a knife, created two. A certain superficial analogy is furnished here by a pair of pendulums which swing together because they are lashed together with string. Cut the string and they swing separately and the number of macroscopic parameters needed to describe the instantaneous state of the system is doubled. Incidentally, one rather eerie feature of these experiments is that one must surely say also that the operation creates two independent consciousnesses where there was only one before.

8.1.2. The Role of the Cortex

It is generally believed that the alterations underlying memory occur, largely at least, in the cortex, although I do not think this could be said to be completely proved. If this is so then the question arises as to whether the cortex is nearer to the limit of being a functionally integrated relatively autonomous unit or to being little more than a static store, in much the sense that a ferrite core store is in a computer, in which items are marked and to which reference is made.

There is some evidence which suggests that the second possibility is nearer to the truth. The histological structure of the cortex suggests a primary organization in vertical chains, i.e. at right angles to the surface (Fulton, 1961; Colonnier, 1966). Also Sperry and Miner's experiments on cutting the cortex seem to imply that integration over long distances within the cortex, i.e. parallel to its surface, is not of great functional significance. In other work, Burns (1958) finds the activity of slabs of cortex is critically dependent on connections coming into the cortex from below, i.e. at right angles to its surface. This is not inconsistent with the idea that the cortex has little intrinsic activity of its own, but the neurones of each little region independently fire in response to signals from elsewhere in the brain. Of course, even if this is approximately true, it must be a great oversimplification of the real situation.

One way in which such a role for the cortex might be integrated into the general activity of the brain is if there is a continuous shuttling of activity between the cortex and subcortical structures. Such an idea was already present in the ideas of Penfield and Jasper (1954), Penfield (1958, p. 8 and Chapter 3), which are currently unfashionable but by no means disproved. If we write the current state of activity of the cortex as C and of the subcortical structure as S, the extreme view of the cortex expressed above might cause us to contemplate representing the shuttling by the equations

$$\dot{C} = f(S)$$
$$\dot{S} = g(C, S), \qquad (1)$$

as a crude first approximation (memory here would reside in alterations of f or g).

8.1.3. Unification of Activity

The question of which bit of the nervous system has overriding control in any given situation has a number of interesting aspects. The first thing to note is that there is considerable hierarchy of structure in a mammalian brain, the senior end of the hierarchy being in the brain and the junior end in the spinal cord. The spinal cord has a great number of intrinsic capabilities (spinal reflexes) which are, or can be, modified by commands from the brain. If it is surgically separated from the rest of the brain then, after a period of recovery which also poses very interesting problems, the isolated spinal cord executes many of the reflexes by itself (Ruch, Patton, Woodbury and Towe, 1964, Chapter 7).

How is the highest level of control organized? Here one hypothetical possibility would be for the hierarchy to converge to a single nerve cell, a dictator for the nervous system. With some invertebrate animals the problem of control may be regarded as operating in this way. For example, the polychaete worm *Myxicola infundibulum*, which lives in the sand at the low-tide mark, sticks a crown of tentacles out in order to respire and feed. If disturbed, it contracts and disappears completely into the sand, returning passively after a certain relaxation period. It seems that this may be the only decision of which it is capable, to disappear or not, and this decision is completely an all-or-none affair (Roberts, 1962). In *Myxicola* there really is convergence on to a single nerve cell; a nerve cell which, appropriate to its dictatorial position, is extremely large (up to 1.7 mm across the axon).

The case of *Myxicola* can be matched in more complicated invertebrates. In crustacea, for example, many functions are performed by one or two nerve cells, which would be performed in vertebrates by very many cells acting in parallel. The lobster and crayfish tails each contain four giant axons and a single pulse in any one of these will initiate the violent defensive snap of the whole tail (Wiersma, 1961, p. 255). However, even then, once our animal is capable of more than one activity the problem arises of decisions between them and the control of such decisions. One would be hard put to it plausibly to assign the power of making many different decisions to a single nerve cell with its all-or-none output.

In fact, to a great extent, animals are only capable of performing one action at a time much as we have great difficulty in thinking of more than one thing at a time. This point was emphasized by Sherrington (1940) who said "Where it is a question of 'mind' the nervous system does not integrate itself by centralization upon one pontifical cell. Rather it elaborates a million-fold democracy whose each unit is a cell." If we accept that the nervous system is a democracy, or even an oligarchy, one immediately has the problem of how are the votes collected and how is the result of the vote communicated as a command to all the nerve cells involved. Eccles (1965, p. 36) has also emphasized this problem saying: "The antithesis must remain that our brain is a democracy of ten thousand million nerve cells, yet it provides us with a unified experience." Of course we must not take the word "democracy" to imply that each nerve cell is likely to have an equal vote, but rather that there is some sort of collective decision-making procedure in which the decision can properly be said to be a property of the aggregate rather than of any particular nerve cell within it.

Finally, note that if we adopt the view of the cortex which is suggested in Section 8.1.2, we are really forced to assign the decision-making role to to the sub-cortical structures, that is if we use equations (1), to the function $g(C, S)$.

8.2. Some Theoretical Problems

8.2.1. The Existence of Invariants of the Motion

The theoretical physicist has discovered the existence of a number of quantities whose values remain indefinitely unchanged for any isolated system. The constancy of such a quantity is known experimentally, in many situations at least, to be true to an astonishing degree of accuracy and the physicist says that the quantity is a "constant or invariant of the motion" or that it satisfies a conservation law. The law of conservation of energy is perhaps the best known, but there are several others, such as the conservation of total angular momentum, of electric charge, of baryon number.

The question naturally arises as to whether there are any aspects of nervous activity which satisfy, even approximately, conservation laws in the absence of input or output and whose alteration in the presence of input and output could be written, in analogy with the first law of thermodynamics, as

$$\Delta A = \Delta I - \Delta O, \tag{2}$$

where A is the content of the quantity in the central nervous system, ΔI is the amount coming in through the sensory input and ΔO the amount going out through, mainly, the motoneurones. The sort of quantity that we are contemplating here is illustrated by taking A equal to the total number of action potentials in the central nervous system. We know, however, from our investigations of random networks in Chapter 5 that the law $\Delta A = 0$ cannot be true for *all* isolated networks of neurones. Are there any such quantities?

A further consideration of a random network of excitatorily connected neurones makes it seem rather unlikely that there are any very useful ones. We shall discuss the point for logical neurones, but a very similar argument could be given for real time neurones. Suppose we have M logical neurones with parameters $n > \theta > 1$ (cf. Section 5.1.1). Then out of all the 2^M possible initial states for the network, all those having (cf. Theorem 5.5) $p > p_0$ pass to the unique state in which every neurone fires, and all starting with $p < p_0$ finish up in the unique inactive state. Just possibly, a few having $p \approx p_0$ remain indefinitely near that intermediate activity. Therefore any conserved quantity A must have one of two values, A_1 say if $p > p_0$ and A_0 if $p < p_0$, for all states of the network except possibly for a few states which have $p \approx p_0$.

This shows the kind of obstacle one is up against. There is, however, a long tradition of talking about "nervous energy" in such a way (e.g. "he is full of nervous energy") as to suggest that it is thought of as a quantity having a conservation law, like physical energy. Indeed, such ideas of conservation are clearly present in much of Freud's work. I think it is correct to say that there is no *a priori* reason to expect such laws to exist for nervous activity and that the concept of nervous energy is not theoretically well-based, at least yet (see also a discussion by Hinde, 1960). The appealing idea that it should be defined as the total number $N(a)$ of action potentials faces the difficulty we have just mentioned.

However, although there are obstacles it may be argued that they are not necessarily insurmountable. This counterargument might run as follows: "It is true that you can construct rather awkward hypothetical examples but these are not naturally-occurring brains. It may well be that in the latter the mode of construction is such as to make some one or more quantities, possibly even your $N(a)$ above, approximately or even quite accurately conserved. If you ask me why such a surprising thing should happen, I can say that there may well be a selective advantage in having a brain which tends to maintain a fairly constant level of activity in the absence of any specific disturbance." Such an attitude is perfectly reasonable although one might perhaps think *a priori* that the maintainance of activity implicit in a stable limit cycle (cf. Sections 7.1 and 7.2.3) might have even more advantage than that due to a conservation law. At any rate, it remains to be decided.

These questions have been discussed from various points of view by Cragg and Temperley (1954, 1955), Elsasser (1962), Griffith (1966b) and Cowan (1968). Cowan obtains a constant of the activity but at the expense of an assumption (his equation (7)) which is certainly not obviously true (note that his equation (3) is in error, ∂v_r should read ∂v_s; and in his equation (6), the first round bracket should have ln in front). Pringle (1951) tried to formulate the nervous system as the union of a number of oscillators, each like real neuronal versions of the closed loops shown in example 4 of Section 3.1.2 or generalizations of them, and Cowan's attempt is along the same lines. The question the reader has to ask himself is whether he thinks the special assumptions made about the properties of neurones and the equations describing their interactions are natural and reasonable ones or whether their only merit is that they can be used to deduce a conservation law (a similar question applies in the different context of the aggregate of control oscillators in cells suggested in Goodwin, 1963). Of course, the truth may be somewhere in between.

8.2.2. Group Theory and the Sensory Input

Slightly away from our general theme in this chapter, I mention briefly here the search for invariant features of the sensory input, which has some mathematical interest. The idea runs as follows. We distinguish usually some objects as different and class others as the same—either as literally indistinguishable or, say if both are circles, as being in the same class. It has been suggested (Pitts and McCulloch, 1947) that we should regard such classifications as group-theoretic in the sense that if A and B are both circles there exists an element g of the group of all translations, rotations and dilations of three-dimensional space which carries A into B. Thus $B = gA$. Such an idea, although attractive, has the following defect. Suppose g represents a contraction of all dimensions by a factor of 2. Then we construct the series $A, gA, g^2A, \ldots, g^nA$, from an original clearly-visible circle A. Ultimately g^nA will be indistinguishable from a point P at its centre. Then there should exist an element h, of the group of *all* operations carrying objects into indistinguishable ones, such that $P = hg^n A$. But then hg^n also belongs to the group, which is a contradiction as A and P are distinguishable. Because of this, it may well be that the concept of tolerance space introduced by Zeeman (1962), which obviates this sort of difficulty, may give a more rewarding approach.

8.2.3. Statistical Neurodynamics

It is clear that for our brains at least, there are two natural levels of description. One talks about detailed knowledge of connections, internal potentials, etc., in the constituent nerve cells. To specify the present state of nervous activity at this level requires at least 10^{10} numbers, probably more. The other level is that of the psychologist, with gross definitions of stimulus or response, or of introspection, the results of which can often be approximated with ordinary language. Furthermore, although we wish to explain the latter level in terms of the former, our real interest lies in the latter and if, for example, there are features of cellular activity irrelevant to behavior then they are also of relatively little interest to us.

The situation is therefore superficially very similar to that which obtains in statistical mechanics, as it applies to the relation between macroscopic thermodynamic quantities and the underlying microscopic description in terms of the complete specification of the states of all the individual atoms or molecules, and the prime requirements for a statistical theory are present (Tolman, 1938, especially Section 1). These are first that we could not, even if we knew all the necessary parameters, actually solve in detail the 10^{10} or more coupled neuronal "equations of motion" necessary to follow the state of the system in detail as a function of time. Second, that there exists a simpler "macroscopic" level of description which is really our main ultimate object of interest so that we do not wish, even if we could, to follow the "microscopic" state in detail but merely wish to use it to understand the time development of the macroscopic state. One most important aspect of this is that we only wish to specify at the macroscopic level the initial conditions of any calculation we may make. This leads immediately to the problem of whether the fundamental assumptions of equal *a priori* probabilities and random *a priori* phases hold for nerve cell aggregates and, if not, whether we can find anything to replace them.

It is important to realize that normal statistical mechanics depends absolutely crucially on these assumptions and the laws of mechanics which may be used to justify them. If they are not true, we still can and should be trying to formulate a statistical neurodynamics but we must then expect that, unless we are remarkably lucky, it will be much more difficult and differ enormously in structure from the statistical mechanics of physicists.

Let us investigate this in relation to an isolated network of M McCulloch–Pitts neurones, for which the issues appear in a particularly clear form. The network has $x = 2^M$ possible different states, which could be numbered by a parameter $i = 1, 2, \ldots, x$. There is a superficial analogy here with the quantum statistical mechanical situation of a set of M subsystems each having two possible quantum states as, for example, a set of M atoms each having spin $\frac{1}{2}$ and no further degeneracy of interest for the problem (cf. Cragg and Temperley, 1954, 1955; Griffith, 1966b).

Suppose the network is a typical randomly connected one with $n > \theta > 1$, as discussed in Sections 5.1.1 to 5.1.3. The natural macroscopic parameter we know for such a network is p, the fraction of neurones firing. We have seen that p normally becomes either 0 or 1, in fact with a relaxation time which is usually little more than one unit τ of time and virtually never more than a few units (cf. equation 5.2). So any isolated system which has been left longer than that will have settled down either into the unique state with $p = 0$ (all neurones inactive) or the unique state with $p = 1$ (all neurones active). Hence statistical considerations would seem almost irrelevant here.

The next simplest situation is a randomly connected network including inhibitory connections and having, say, one stable intermediate level of activity given by a value p_0 for the macroscopic parameter p. Suppose we allow the network to settle down and find that the macroscopic parameter continually and endlessly fluctuates about the value p_0. Can we say anything about the probabilities of the various microscopic descriptions consistent with this?

First, note that each of the 2^M states i of the network has a definite successor in time, which we shall write $S(i)$, which will follow it on any occasion that i occurs. This, of course, is true for any isolated network of McCulloch–Pitts neurones without habituation. Hence, if we start at one state, i_0 say, we can construct indefinitely the sequence of its successors:

$$i_0 \rightarrow i_i = S(i_0) \rightarrow i_2 = S(i_1) \rightarrow \ldots. \quad (3)$$

Ultimately, as there are only 2^M states altogether, a state must occur which was already present earlier in the sequence. That means there must be integers m and $q > m$ such that $i_m = i_q$. Let q be the smallest integer for which this happens. Then all the states $i_m, i_{m+1}, \ldots, i_{q-1}$ are different from each other and the network cycles through these $q-m$ states indefinitely. We shall call the initial m states $i_0, i_1, \ldots, i_{m-1}$ transient states and the remaining $q-m$ states terminal states. Evidently, no matter how it starts, any isolated system ultimately finishes up in a terminal cycle made up entirely of terminal states. We shall call $q-m$ the order of the cycle and we may, of course, have $q-m = 1$ in which case we should say the system has finished up in a single absorbing state. The two unique states corresponding to $p = 0$ and $p = 1$ are very often each absorbing states for randomly connected networks as we saw in Chapter 5.

Evidently, no matter how we start it, an isolated network will ultimately finish up in a terminal cycle. In the particular case that we are now considering there are three natural possibilities for this terminal cycle, namely the state $p = 0$, the state $p = 1$ or a cycle with p fluctuating around $p = p_0$. Other possibilities are not excluded rigorously but are presumably rare. The questions then arise of how many terminal states there are altogether, what are their relative probabilities assuming the initial state is chosen at random from amongst the 2^M states, what is the typical size of m and q, etc? Presumably the number of initial states which finish up at $p = 0$ is roughly given by

$$\sum_{\mu=0}^{(Mp_1)} \binom{M}{\mu},$$

where (Mp_1) is the nearest integer to Mp_1 and p_1 is the lowest unstable stationary point (near θ/n_e, cf. equation (5.19)). Similarly for $p = 1$. Finally, those which finish up near $p = p_0$ are the remainder.

So the problem relates to the states which finish up in terminal cycles around $p = p_0$. Let there be x of these. We now meet what is probably the fundamental difference between the neural and the quantum situation. In the latter the transition probabilities between two states ψ_0 and ψ_1 under the influence of a perturbation V responsible for the transition are equal in both directions because they are proportional respectively to the two sides of the equation (see Dirac, 1947, Section 44)

$$|\langle\bar{\psi}_0|V|\psi_1\rangle|^2 = |\langle\bar{\psi}_1|V|\psi_0\rangle|^2$$

which is true because V must be Hermitian. In the former, it is only very exceptionally that the corresponding thing is true, i.e. that both $i_1 = S(i_0)$ and $i_0 = S(i_1)$. We do not have microscopic reversibility. In fact, apart from the natural tendency of the macroscopic parameter p to move near to p_0, there would not seem to be any very obvious reason for the successor function S to show any particular symmetry or other simplifying structure.

This suggests that we may get a good feel for the situation if we analyse the following simpler situation, which has the advantage of giving us definite and quite simple formulae for all the main quantities of interest. We consider a set of x numbers from 1 up to x and suppose that the successor function is obtained by choosing for each state a successor entirely randomly from the numbers $1, 2, \ldots, x$, i.e. with probability x^{-1} for each. This is done independently for the successor of each number. There are thus x^x different possible successor functions, each having a probability x^{-x} of being chosen.

We now ask what is the mathematical expectation of the number of transient states, of cycles of particular sizes, etc.* We define the following quantities:

Y_x = expected number of terminal elements,
y_x^i = expected number in cycles of order i,
Z_x = expected total number of cycles,
z_x^i = expected number of cycles of order i.

Evidently, from these definitions:

$$y_x^i = iz_x^i,$$
$$Y_x = \sum_{i=1}^{x} y_x^i,$$
$$Z_x = \sum_{i=1}^{x} z_x^i. \tag{4}$$

We first calculate y_x^i. Choose one of the x numbers at random. The probability that it is part of a cycle of order i is given by

$$p^i = \frac{x-1}{x} \cdot \frac{x-2}{x} \cdot \ldots \cdot \frac{x-i+1}{x} \cdot \frac{1}{x} \tag{5}$$

because all of its first $(i-1)$ successors and itself must be different and then finally the cycle must close. Therefore the expectation of the total number in such cycles is

$$y_x^i = xp_x^i = \frac{(x-1)!}{x^{i-1}(x-i)!}. \tag{6}$$

Note that when x is large, $y_x^i \approx 1$ for all small values of i, so on average each system has one absorbing state. Putting equation (6) into formulae (4) gives us immediately expressions for Y_x and Z_x. Some values computed from those formulae are shown in Table 8.1.

TABLE 8.1

Computed values of Y_x, Z_x and $Z_x - \frac{1}{2}\log_e x$

x	Y_x	Z_x	$Z_x - \frac{1}{2}\log_e x$
10	3.66022	1.91303	0.7617
10^2	12.2100	2.97899	0.6764
10^3	39.3032	4.10222	0.6483
10^4	124.999	5.24452	0.6394
10^5	396.000	6.39297	0.6365
10^6	1252.98	7.54335	0.6356

* Many of the mathematical results given here were proved previously by Kruskal (1954), Rubin and Sitgreaves (unpublished results, 1954), Harris (1960) and Riordan (1962).

The total number of states of a McCulloch–Pitts network of M neurones is 2^M so, in any use of our results for discussing the McCulloch–Pitts approximation to a real brain or any substantial part thereof, we shall only be interested in extremely large values for x. Hence we wish to obtain asymptotically correct formulae for Y_x and Z_x as $x \to \infty$. We now do this, using Stirling's series for the Gamma function $\Gamma(x) = (x-1)!$ (see e.g. Whittaker and Watson, 1952, p. 253) to approximate the factorials occurring in equation (6). We obtain

$$\begin{aligned} y_x^i &= \frac{\Gamma(x)}{x^{i-1}\Gamma(x-i+1)} \\ &= \frac{e^{-x}\, x^{x-\frac{1}{2}}(2\pi)^{\frac{1}{2}}(1+1/(12x)+O(x^{-2}))}{x^{i-1}\, e^{-x+i-1}\,(x-i+1)^{x-i+\frac{1}{2}}(2\pi)^{\frac{1}{2}}(1+(x-i+1)^{-1}/(12)+O(x^{-2}))} \\ &= \frac{1+O(ix^{-2})}{e^{i-1}\,\{1-(i-1)x^{-1}\}^{x-i+\frac{1}{2}}}, \end{aligned}$$

whence by taking logarithms and expanding in powers of $(i-1)x^{-1}$ we obtain after some straightforward manipulation

$$\log_e y_x^i = -\frac{\frac{1}{2}i(i-1)}{x} - \frac{i^3}{6x^2} + O(i^2x^{-2}). \tag{7}$$

The quantity y_x^i is a monotonic decreasing function of i for fixed x and using the Cauchy–Maclaurin comparison theorem between series and integrals (Hardy, Littlewood and Pólya, 1952, p. 110) we get the inequalities

$$y_x^x + \int_1^x y_x^\varepsilon\, d\varepsilon < Y_x < 1 + \int_1^x y_x^\varepsilon\, d\varepsilon.$$

Again using Stirling's formula, $y_x^x \doteqdot e^{-x}(2\pi x)^{\frac{1}{2}}$ for large x, and after a certain amount of manipulation of the integrals, using equation (7) for y_x^ε, this yields

$$(\tfrac{1}{2}\pi x)^{\frac{1}{2}} - \tfrac{5}{6} + O(x^{-\frac{1}{2}}) < Y_x < (\tfrac{1}{2}\pi x)^{\frac{1}{2}} + \tfrac{1}{6} + O(x^{-\frac{1}{2}}). \tag{8}$$

Actually we always have $y_x^{i+1} - y_x^i$ small, of the order $1/x$ or less, and this enables us to get a better approximation to Y_x by using the relation

$$\int_i^{i+1} y_x^\varepsilon\, d\varepsilon \doteqdot \tfrac{1}{2}(y_x^i + y_x^{i+1}),$$

which is true to $O(i^2x^{-2})$. As a consequence we obtain the asymptotically correct expression

$$\begin{aligned} Y_x &\doteqdot \int_1^x y_x^\varepsilon\, d\varepsilon - \sum_{i=1}^{x-1} \tfrac{1}{2}(y_x^i - y_x^{i+1}) + 1 \\ &= \int_1^x y_x^\varepsilon\, d\varepsilon - \tfrac{1}{2} + \tfrac{1}{2}y_x^x + 1 \\ &= (\tfrac{1}{2}\pi x)^{\frac{1}{2}} - \tfrac{1}{3}. \end{aligned} \tag{9}$$

As $\sqrt{\frac{1}{2}\pi} = 1.2533$ and $\sqrt{5\pi} = 3.9633$, it is apparent from Table 8.1 that equation (9) yields an approximation to Y_x which is good even for $x = 10$ and is indistinguishable from the calculated value for $x = 10^6$.

The calculation for Z_x proceeds in a similar manner except that $z_x^{i+1} - z_x^i$ is no longer necessarily small (for low i, $z_x^i \doteqdot 1/i$). However, if we replace z_x^i with $z_x^i - i^{-1}$ we obtain a sequence with the desired properties. Hence we can treat Z_x in a similar manner to Y_x providing we make a preliminary manipulation:

$$\begin{aligned} Z_x &= \sum_{\varepsilon=1}^{x} \varepsilon^{-1} + \sum_{\varepsilon=1}^{x} (z_x^{\varepsilon} - \varepsilon^{-1}) \\ &= \log_e x + \gamma + \int_1^x (z_x^{\varepsilon} - \varepsilon^{-1})\, d\varepsilon \\ &= \gamma + \int_1^x z_x^{\varepsilon}\, d\varepsilon, \end{aligned}$$

where $\gamma = 0.5772 \ldots$ is Euler's constant. After a rather lengthy but essentially straightforward expansion of the exponential and treatment of the resultant integral we finally finish up with

$$Z = A + \tfrac{1}{2} \log_e x + O(x^{-\frac{1}{2}}), \tag{10}$$

where

$$A = \tfrac{1}{2}(\gamma + \log_e 2) = 0.635181 \tag{11}$$

We see from Table 8.1 that for $x = 10^6$, Z_x is already quite close to being given correctly by the first two terms in equation (10).

Our results show that, when x is very large, the expectation for the fraction of all the states which are terminal states, which is $x^{-1}Y_x$, is approximately $(\pi/2x)^{\frac{1}{2}}$, in other words a very small proportion of the total. This appears to present an obstacle to any assumption of equal *a priori* probabilities because it suggests that, with states of McCulloch–Pitts neurones also, there may well be an inherent tendency for the system to finish up in one of a minute fraction of the total number of states consistent with a given macroscopic specification, in complete contrast with the situation which holds in quantum statistical mechanics.

However, although the fraction of states, which are in terminal cycles, is small, the mean number is still extremely large. Also the average cycle size, which is Y_x/Z_x, is approximately $(2\pi x)^{\frac{1}{2}}/\log_e x$ and is therefore also extremely large. In fact the same is true of almost all relaxation times for the system as we now see.

Let us start with a state at random and ask how long is the path from this state, through transient states, until we first meet a terminal state. The probability that after j transient steps (including the state we start with) we have just started on a cycle of the order i is

$$p^{ji} = \frac{x-1}{x} \cdot \frac{x-2}{x} \cdot \frac{x-3}{x} \cdot \ldots \frac{x-i-j+1}{x} \cdot \frac{1}{x}, \quad 1 \leqslant i+j \leqslant x. \tag{12}$$

We shall call j the path length. When $j = 0$, we get equation (5). The sum of all the p^{ji} is

$$\sum_{i,j} p^{ji} = \sum_{k=1}^{x} \sum_{i+j=k} p^{ji} = \sum_{k=1}^{x} kp^k,$$

where p^k is given by equation (5). Then

$$kp^k = \frac{(x-1)!}{x^k(x-k)!}\{x-(x-k)\},$$

whence

$$\sum_{i,j} p^{ji} = \sum_k kp^k = \sum_{k=1}^{x} \frac{(x-1)!}{x^{k-1}(x-k)!} - \sum_{k=1}^{x-1} \frac{(x-1)!}{x^k(x-k-1)!} = 1, \tag{13}$$

because all terms except one cancel out in pairs.

The probability that a state chosen at random gives a path of length j before going into some cycle is

$$p_j = \sum_{i=1}^{x-j} p^{ji} = \frac{1}{x} \sum_{k=j+1}^{x} y_x^k. \tag{14}$$

Of course $p_0 = x^{-1}Y_x$. The mean path length is given by

$$L_x = \sum_{i,j} jp^{ji}. \tag{15}$$

which is also the mean relaxation time in McCulloch–Pitts time units τ for the system to pass into a terminal cycle. We simplify this by noting that

$$\sum_{i+j=k} jp^{ji} = p^k \sum_{j=0}^{k-1} j = \tfrac{1}{2}k(k-1)p^k,$$

whence

$$L_x = \tfrac{1}{2} \sum_{k=1}^{x} k(k-1)p^k$$
$$= \frac{1}{2x} \sum_{k=1}^{x} k^2 y_x^k - \tfrac{1}{2},$$

where we used equations (6) and (13). Again we can deduce an asymptotic expression for L_x. We are now summing a function of the form

$$f(k) = k^2 + \alpha k + \beta + yk^{-1} + \ldots$$

for the early members of the series. $f(1) = 1$. Define $f(k) = 1$ for $0 \leqslant k \leqslant 1$. Then if we approximate the sum by Simpson's rule, which is exact for $k^2 + \alpha k + \beta$, we shall get rid of the errors arising from the large jumps in

$f(k)$ for low k. Thus:

$$\int_0^x f(k)\,dk \approx \tfrac{1}{3}\{f(0)+4f(1)+f(2)\}+\tfrac{1}{3}\{f(2)+4f(3)+f(4)\}+\ldots$$
$$= \tfrac{4}{3}\sum f(2\lambda+1)+\tfrac{2}{3}\sum f(2\lambda).$$

Also

$$\int_0^x f(k)\,dk = \int_0^1 f(k)\,dk + \int_1^x f(k)\,dk$$
$$\approx 1+\tfrac{4}{3}\sum f(2\lambda)+\tfrac{2}{3}\sum f(2\lambda+1).$$

So on addition we get

$$\int_0^x f(k)\,dk \approx \tfrac{1}{2} + \sum_{k=1}^{x} f(k).$$

After this we simplify the integral as we did for Y_x and Z_x to get finally for the asymptotically correct expression for mean path length

$$L_x = (\tfrac{1}{8}\pi x)^{\frac{1}{2}} - \tfrac{2}{3}. \tag{16}$$

For a McCulloch–Pitts network of M neurones we have 2^M states altogether and therefore are obviously interested in systems with random successors in which x may be a sizeable fraction of 2^M. Even for quite small M, L_x is then very large. For example, if $x = 2^{100}$ and the time intervals τ are 1 msec, then $L_x \approx$ 20 000 years. For appreciably larger exponents L_x becomes more than astronomical; for networks with as many neurones as the human brain L_x is of the order of 2^{10^9} years or more.

We have discussed an approximation to the McCulloch–Pitts situation, because we can thereby get definite formulae. Using M real time neurones we should presumably construct some sort of M-dimensional phase space for the system, probably based on the M values of $V(t)$ for the neurones. The following kind of question then arises. If an outside observer only knows that I am thinking about M-dimensional phase space (a macroscopic characterization of the state of my brain) would he be justified in acting on the assumption that all regions in that M-dimensional phase space which refers to my brain and which contain only points which are consistent with the macroscopic characterization, have probabilities in proportion to their extension, and trying to calculate my future thoughts on this basis? The McCulloch–Pitts results show that if I am left to think about the topic for long enough, the answer is probably no, but if the enormously long relaxation times apply also to the real neurone situation, which is not obvious, the assumption of equal *a priori* probabilities could turn out not to be so bad after all because the system may never have time to relax significantly towards its terminal non-uniform condition.

APPENDIX

Lagrange's Method of Undetermined Multipliers

For those readers not familiar with this method, we prove it in a form sufficient for our application in Section 6.3. Given a function H of n variables $\phi_1, \phi_2, \ldots, \phi_n$ the stationary values of this function are obtained by equating to zero the coefficients of each of the $\delta\phi_i$ in the equation

$$\delta \mathrm{H} = \sum_{i=1}^{n} \frac{\partial \mathrm{H}}{\partial \phi_i} \delta\phi_i = 0, \tag{1}$$

i.e. they are given by solving the n simultaneous equations

$$\frac{\partial \mathrm{H}}{\partial \phi_1} = \frac{\partial \mathrm{H}}{\partial \phi_2} = \ldots = \frac{\partial \mathrm{H}}{\partial \phi_n} = 0. \tag{2}$$

However, if the ϕ_i are only allowed to vary on the hyperplane

$$\sum_{i=1}^{n} \phi_i = 1, \tag{3}$$

this means that only small variations $\delta\phi_i$ satisfying

$$\sum_{i=1}^{n} \delta\phi_i = 0 \tag{4}$$

are allowed. This is just one extra relation between the n quantities $\delta\phi_i$ and so therefore $\delta\phi_1, \delta\phi_2, \ldots, \delta\phi_{n-1}$ can still be chosen arbitrarily providing we then put

$$\delta\phi_n = -\sum_{i=1}^{n-1} \delta\phi_i. \tag{5}$$

Substituting this into equation (1) we get

$$\sum_{i=1}^{n-1} \left(\frac{\partial \mathrm{H}}{\partial \phi_i} - \frac{\partial \mathrm{H}}{\partial \phi_n} \right) \delta\phi_i = 0 \tag{6}$$

and as the $\delta\phi_i$ in equation (6) can be chosen arbitrarily it follows that the $(n-1)$ coefficients must all be zero. Therefore in place of equation (2) we have the weaker set of equations

$$\frac{\partial \mathrm{H}}{\partial \phi_1} = \frac{\partial \mathrm{H}}{\partial \phi_2} = \ldots = \frac{\partial \mathrm{H}}{\partial \phi_n}, \tag{7}$$

which, together with equation (3), give n conditions to determine the values of $\phi_1, \phi_2, \ldots, \phi_n$ at the stationary point.

Lagrange's method is equivalent to the preceding one but treats $\delta\phi_1, \delta\phi_2, \ldots, \delta\phi_n$ symmetrically by introducing an extra parameter μ. It is clear that

$$\delta \mathrm{H} - \mu\delta \sum \phi_i = 0 \tag{8}$$

for any μ, but in Lagrange's method μ is determined by requiring that the $\delta\phi_i$ in equation (8) should all have their coefficients equal to zero. This is true if $\partial \mathrm{H}/\partial\phi_i = \mu$ for every i which, after elimination of μ, is the same as equations (7). For a more general discussion of this method, see Courant (1936).

REFERENCES

Agranoff, B. W., Davis, R. E. and Brink, J. J. (1966). Chemical studies on memory fixation in goldfish. *Brain Research*, **1,** 303–309.

Allanson, J. T. (1956). Some properties of a randomly connected neural network. *In* "Information Theory" (Ed. C. Cherry), pp. 303–312. Academic Press, New York and London.

Anderson, P. and Andersson, S. A. (1968). "Physiological Basis of the Alpha Rhythm". Appleton Century Crofts, New York.

Andjus, R. K., Knöpfelmacher, F., Russell, R. W. and Smith, A. U. (1956). Some effects of severe hypothermia on learning and retention. *Q. Jl. exp. Psychol.* **8,** 15–23.

Andronov, A. A., Vitt, A. A. and Khaikin, S. E. (1966). "Theory of Oscillators". Pergamon Press, Oxford.

Arbib, M. A. (1965). "Brains, Machines and Mathematics". McGraw-Hill, New York.

Arden, G. B. and Söderberg, U. (1961). The transfer of optic information through the lateral geniculate body of the rabbit. *In* "Sensory Communication" (Ed. W. A. Rosenblith), pp. 521–544. The M.I.T. Press, John Wiley, New York.

Ashby, W. R., Von Foerster, H. and Walker, C. C. (1962). Instability of pulse activity in a net with threshold. *Nature, Lond.* **196,** 561–562.

Bailey, N. T. J. (1964). "The Elements of Stochastic Processes". John Wiley, New York.

Barlow, H. B., Hill, R. M. and Levick, W. R. (1964). Retinal ganglion cells responding selectively to direction and speed of image motion in the rabbit. *J. Physiol.* **173,** 377–407.

Barondes, S. H. and Cohen, H. D. (1966). Puromycin effect on successive phases of memory storage. *Science*, **151,** 594–595.

Bass, L. (1964). Potential of liquid junctions. *Trans. Faraday Soc.* **60,** 1914–1919.

Bass, L. (1965). Theory of liquid junction potentials. *Proc. Phys. Soc.* **85,** 1045–1046.

Bass, L. and Moore, W. J. (1967). Electric fields in perfused nerves. *Nature, Lond.* **214,** 393–394.

Basson, A. H. and O'Connor, D. J. (1965). "Introduction to Symbolic Logic". University Tutorial Press, London.

Bateman Manuscript Project (1953). McGraw-Hill, New York.

Bellman, R. and Cooke, K. L. (1963). "Differential-Difference Equations". Academic Press, New York and London.

Betz, A. (1968). Oscillatory control of glycolysis as a model for biological timing processes. *In* "Quantitative Biology of Metabolism" (Ed. A. Locker), pp. 205–216. Springer-Verlag, Berlin.

Beurle, R. L. (1956). Properties of a mass of cells capable of regenerating pulses. *Phil. Trans. R. Soc.* **A240,** 55–94.

Blinkov, S. M. and Glezer, I. T. (1968). "The Human Brain in Figures and Tables". Basic Books, Plenum Press, New York.

Bodian, D. (1962). The generalized vertebrate neuron. *Science, N.Y.* **137,** 323–326.

Bodian, D. (1967). Neurons, circuits and neuroglia. *In* "The Neurosciences" (Eds G. C. Quarton, T. Melnechuk and F. O. Schmitt), pp. 6–24. Rockefeller University Press, New York.

Bogoch, S. (1968). "The Biochemistry of Memory". Oxford University Press, London.

Bogoch, S., Editor (1969). "The Future of the Brain Sciences". Plenum Press, New York.

Boyd, I. A. and Martin, A. R. (1956). The end-plate potential in mammalian muscle. *J. Physiol., Lond.* **132,** 74–91.

Brillouin, L. (1962). "Science and Information Theory". Academic Press, New York and London.

Brodal, A. (1969). "Neurological Anatomy in Relation to Clinical Medicine". Oxford University Press, London.

Brown, D. D. and Dawid, I. B. (1968). Specific gene amplification in oocytes. *Science, N.Y.* **160,** 272–280.

Bruesch, S. R. and Arey, L. B. (1942). The number of myelinated and unmyelinated fibers in the optic nerve of vertebrates. *J. comp. Neurol.* **77,** 631–665.

Buchthal, F. and Rosenfalck, A. (1966). Evoked action potentials and conduction velocity in human sensory nerves. *Brain Res., Osaka* **3,** 1–122.

Burns, B. D. (1958). "The Mammalian Cerebral Cortex". Edward Arnold, London.

Burns, B. D. and Salmoiraghi, G. C. (1960). Repetitive firing of respiratory neurones during their burst activity. *J. Neurophysiol.* **23,** 27–46.

Caianiello, E. R. (1961). Outline of a theory of thought processes and thinking machines. *J. theor. Biol.* **1,** 204–235.

Caianiello, E. R., de Luca, A. and Ricciardi, L. M. (1968). Neural networks: Reverberations, constants of motion, general behaviour. *In* "Neural Networks" (Ed. E. R. Caianiello), pp. 92–99. Springer-Verlag, Berlin.

Cajal, R. Y. (1952). "Histologie du Système Nerveux". 2 Vol. Istituto Ramon y Cajal, Madrid.

Carslaw, H. S. and Jaeger, J. C. (1959). "Conduction of Heat in Solids". Clarendon Press, Oxford.

Castillo, J. D. and Katz, B. (1954). Quantal components of the end-plate potential. *J. Physiol., Lond.* **124,** 560–573.

Chance, B., Estabrook, R. W. and Ghosh, A. (1964). Damped sinusoidal oscillations of cytoplasmic reduced pyridine nucleotide in yeast cells. *Proc. natn. Acad. Sci. U.S.A.* **51,** 1244–1251.

Chandrasekhar, S. (1943). Stochastic problems in physics and astronomy. *Rev. mod. Phys.* **15,** 1–89.

Cobb, S. (1965). Brain size. *Archs Neurol. Chicago* **12,** 555–561.

Cole, K. S. (1968). "Membranes, Ions and Impulses". University of California Press, Berkeley and Los Angeles.

Colonnier, M. L. (1966). Structural design of the neocortex. *In* "Brain and Conscious Experience" (Ed. J. C. Eccles), pp. 1–23. Springer-Verlag, Berlin.

Courant, R. (1936). "Differential and Integral Calculus", Vol. 2, pp. 190–198. Blackie and Son, London.

Cowan, J. D. (1968). Statistical mechanics of nervous nets. *In* "Neural Networks" (Ed. E. R. Caianiello), pp. 181–188. Springer-Verlag, Berlin.

Cowan, J. and Winograd, S. (1963). "Reliable Computation in the Presence of Noise". M.I.T. Press, Cambridge, Mass.

Cox, D. R. (1962). "Renewal Theory". Methuen, London.

Cox, D. R. and Miller, H. D. (1965). "Theory of Stochastic Processes". John Wiley, New York.

Cragg, B. G. (1967). The density of synapses and neurones in the motor and visual areas of the cerebral cortex. *J. Anat.* **101**, 639–654.

Cragg, B. G. (1968). Are there structural alterations in synapses related to functioning? *Proc. Roy. Soc.* B. **171**, 319–324.

Cragg, B. G. and Temperley, H. N. V. (1954). The organization of neurones: a co-operative analogy. *Electroenceph. clin. Neurophysiol.* **6**, 85–92.

Cragg, B. G. and Temperley, H. N. V. (1955). Memory: the analogy with ferromagnetic hysteresis. *Brain* **78**, 304–316.

Cramér, H. (1955). "The Elements of Probability Theory and some of its Applications". Wiley, New York.

Creutzfeldt, O., Fuster, J. M., Herz, A. and Straschill, M. (1966). Information transmission in the visual system. *In* "Brain and Conscious Experience" (Ed. J. C. Eccles), pp. 138–164.

Crile, G .W. and Quiring, D. P. (1940). A record of the body weight and certain organ and gland weights of 3690 animals. *Ohio J. Sci.* **40**, 219–259.

Davson, H. (1967). "Physiology of the Cerebrospinal Fluid". J. and A. Churchill, London.

de Robertis, E. D. P. and Carrea, R. (1965). Biology of Neuroglia: Progress in Brain Research", Vol. 15. Elsevier, Amsterdam.

Denbigh, K. G., Hicks, M. and Page, F. M. (1948). The kinetics of open reaction systems. *Trans. Faraday Soc.* **44**, 479–494.

Deutsch, J. A. (1962). Higher nervous functions: the physiological bases of memory. *A. Rev. Physiol.* **24**, 259–286.

Dirac, P. A. M. (1947). "The Principles of Quantum Mechanics". Clarendon Press, Oxford.

Dixon, K. C. (1967). Neuronal membrane patterns in memory and concussion. *Lancet*, 7 January, pp. 27–28.

Donaldson, P. E. K. (1958). "Electronic Apparatus for Biological Research". Butterworth, London.

Eccles, J. C. (1953). "The Neurophysiological Basis of Mind". Oxford University Press, London.

Eccles, J. C. (1960). "The Physiology of Nerve Cells". Oxford University Press, London.

Eccles, J. C. (1964). "The Physiology of Synapses". Academic Press, New York and London.

Eccles, J. C. (1965). "The Brain and the Unity of Conscious Experience." Cambridge University Press, London.

Eccles, J. C., Ito, M. and Szentágothai, J. (1967). "The Cerebellum as a Neuronal Machine". Springer-Verlag, Berlin.

Eccles, J. C. and Jaeger, J. C. (1958). The relationship between the mode of operation and the dimensions of the junctional regions at synapses and motor end-organs. *Proc. R. Soc.* B. **148**, 38–56.

Elias, H. and Schwartz, D. (1969). Surface areas of the cerebral cortex of mammals determined by stereological methods. *Science, N.Y.* **166,** 111–113.

Elsasser, W. M. (1962). Physical aspects of non-mechanistic biological theory. *J. theor Biol.* **3,** 164–191.

Farley, B. G. and Clark, W. A. (1961). Activity in networks of neuron-like elements. *In* "Information Theory, 4th London Symposium" (Ed. C. Cherry), pp. 242–248.

Feder, W. (1968). Editor: *Ann. N.Y. Acad. Sci.* **148.** Entire volume on Bioelectrodes.

Fitzhugh, R. (1961). Impulses and physiological states in theoretical models of nerve membrane. *Biophys. J.* **1,** 445–466.

Fulton, J. F. (1961). "Physiology of the Nervous System", p. 305. Oxford University Press, London.

Galambos, R. (1961). A glia-neural theory of brain function. *Proc. natn. Acad. Sci. U.S.A.* **47,** 129–136.

Gallager, R. G. (1968). "Information Theory and Reliable Communication". John Wiley, New York.

Gerschenfeld, H. M., Ascher, P. and Tauc, L. (1967). Two different excitatory transmitters acting on a single molluscan neurone. *Nature, Lond.* **213,** 358–359.

Gerstein, G. L. and Mandelbrot, B. (1964). Random walk models for the spike activity of a single neuron. *Biophys. J.* **4,** 41–68.

Glassman, E. (1969). Biochemistry of learning: An evaluation of the role of RNA and protein. *A. Rev. Biochem.* **38,** 605–646.

Glasstone, S. (1946). "Textbook of Physical Chemistry". D. Van Nostrand, New York.

Glees, P. (1955). "Neuroglia: Morphology and Function". C. C. Thomas, Springfield, Ill.

Glees, P. (1958). The biology of the neuroglia: a summary. *In* "Biology of Neuroglia" (Ed. W. F. Windle), pp. 234–242.

Gluss, B. (1967). A model for neuron firing with exponential decay of potential resulting in diffusion equations for probability density. *Bull. math. Biophys.* **29,** 233–243.

Gnedenko, B. V. (1962). "The Theory of Probability". Chelsea Publishing Company, New York.

Goldman, D. (1943). Potential, impedance and rectification in membranes. *J. gen. Physiol.* **27,** 37–60.

Good, I. J. (1966). Speculations concerning the first ultraintelligent machine. *Adv. Comput.* (1965), **6,** 31–88.

Goodwin, B. C. (1963). "Temporal Organization in Cells". Academic Press, London and New York.

Grey Walter, W. (1953). "The Living Brain". W. W. Norton, New York.

Griffith, J. S. (1963a). On the stability of brain-like structures. *Biophys. J.* **3,** 299–308.

Griffith, J. S. (1963b). A field theory of neural nets: I. Derivation of field equations. *Bull. math. Biophys.* **25,** 111–120.

Griffith, J. S. (1964). "The Theory of Transition-metal Ions. Cambridge University Press, London.

Griffith, J. S. (1965a). A field theory of neural nets. II. Properties of the field equations. *Bull. math. Biophys.* **27,** 187–195.

Griffith, J. S. (1965b). Information theory and memory. *In* "Molecular Biophysics" (Eds B. Pullman and M. Weissbluth), pp. 411–435. Academic Press, New York and London.

Griffith, J. S. (1966a). A theory of the nature of memory. *Nature, Lond.* **211,** 1160–1163.

Griffith, J. S. (1966b). Modelling neural networks by aggregates of interacting spins. *Proc. Roy. Soc.* A. **295,** 350–354.

Griffith, J. S. (1967a). "A View of the Brain". Oxford University Press, London.

Griffith, J. S. (1967b). Neural organization underlying conscious thought. *Nature, Lond.* **214,** 345–349.

Griffith, J. S. (1967c). "The Neural Basis of Conscious Decision". Bedford College, London.

Griffith, J. S. (1967d). Brains and computers. *Theoria to Theory,* **2,** 39–46.

Griffith, J. S. (1968a). Mathematics of cellular control processes. Part I: Negative feedback to one gene; Part 2: Positive feedback to one gene. *J. theor. Biol.* **20,** 202–208, 209–216.

Griffith, J. S. (1968b). The unification of neural activity. *Proc. Roy. Soc.* B. **171,** 353–359.

Griffith, J. S. (1968c). Memory and cellular control processes. *In* "Quantitative Biology of Metabloism" (Ed. A. Locker), pp. 234–244. Springer-Verlag, Berlin.

Griffith, J. S. (1970). The transition from short to long term memory. *In* "Short-term Changes in Neural Activity and Behaviour" (Eds R. A. Hinde and G. Horn), pp. 499–516. Cambridge University Press, London.

Griffith, J. S. and Horn, G. (1963). Functional coupling between cells in the visual cortex of the unrestrained cat. *Nature, Lond.* **199,** 876, 893–895.

Griffith, J. S. and Horn, G. (1966). An analysis of spontaneous impulse activity of units in the striate cortex of unrestrained cats. *J. Physiol., Lond.* **186,** 516–534.

Griffith, J. S. and Mahler, H. R. (1969). A DNA ticketing theory of memory. *Nature, Lond.* **223,** 580–582.

Guggenheim, E. A. (1965). Liquid junction potentials. *Proc. phys. Soc.* **85,** 393–394.

Hahn, W. (1963). "Theory and Application of Liapunov's Direct Method". Prentice-Hall, New York.

Hardy, G. H., Littlewood, J. E. and Pólya, G. (1952). "Inequalities". Cambridge University Press, London.

Harmon, L. D. (1959). Artificial neurone. *Science, N.Y.* **129,** 962–963.

Harmon, L. D. (1961). Studies with artificial neurons: I. Properties and functions of an artificial neuron. *Kybernetik,* **1,** 89–101.

Harmon, L. D. and Lewis, E. R. (1966). Neural modeling. *Physiol. Rev.* **46,** 513–591.

Harris, B. (1960). Probability distributions related to random mappings. *Ann. math. Stat.* **31,** 1045–1062.

Hebb, C. (1970). CNS at the cellular level: Identity of transmitter agents. *A. Rev. Physiol.* **32,** 165–192.

Hebb, D. O. (1961). "Organization of Behavior". Science Editions, New York.

Herman, C. J. and Lapham, L. W. (1969). Neuronal polyploidy and nuclear volumes in the cat central nervous system. *Brain Res., Osaka* **15,** 35–48.

Herz, A. Creutzfeldt, O. and Fuster, J. (1964). Statistische Eigenschaften der Neuronaktivität in ascendierenden visuellen System. *Kybernetik,* **2,** 61–71.

Higgins, J. (1967). The theory of oscillating reactions. *Ind. Engng Chem.* **59,** 18–62.

Hinde, R. A. (1960). Energy models of motivation. *Symp. Soc. exp. Biol.* **14,** 199–213.

Hodgkin, A. L. (1965). "The Conduction of the Nervous Impulse". University Press, Liverpool.

Hodgkin, A. L. and Huxley, A. F. (1952). A quantitative description of membrane current and its application to conduction and excitation in nerve. *J. Physiol.* **117,** 500–544.

Holloway, R. L. (1966). Cranial capacity and neuron number: A critique and proposal. *Am. J. phys. Anthrop.* **25,** 305–314.

Horn, G. (1962). Regular impulse activity of single units in the cat striate cortex. *Nature, Lond.* **194,** 1084–1085.

Horn, G. (1967). Neuronal mechanisms of habituation. *Nature, Lond.* **215,** 707–711.

Horridge, G. A. (1968). "Interneurones". W. H. Freeman, London.

Hubel, D. H. and Wiesel, T. N. (1959). Receptive fields of single neurons in the cat's striate cortex. *J. Physiol., Lond.* **148,** 574–591.

Hubel, D. H. and Wiesel, T. N. (1965). Receptive fields and functional architecture in two non-striate visual areas (18 and 19) of the cat. *J. Neurophysiol.* **28,** 229–289.

Hydén, H. (1960). The neuron. *In* "The Cell" (Eds J. Brachet and A. E. Mirsky), Vol. 4, pp. 215–323. Academic Press, New York and London.

Hydén, H. (1962). A molecular basis of neuron-glia interaction. *In* "Macromolecular Specificity and Biological Memory" (Ed. F. O. Schmitt), pp. 55–69. M.I.T. Press, Cambridge, Mass.

Ingram, V. M. (1966). "The Biosynthesis of Macromolecules". W. A. Benjamin, New York.

Johannesma, P. I. M. (1968). Diffusion models for the stochastic activity of neurons. *In* "Neural Networks" (Ed. E. R. Caianiello), pp. 116–144. Springer-Verlag, Berlin.

John, E. R. (1967). "Mechanisms of Memory". Academic Press, New York and London.

Karlsson, U. (1966). Three-dimensional studies of neurons in the lateral geniculate nucleus of the rat. *J. Ultrastruct. Res.* **16,** 482–504.

Katz, B. (1966). "Nerve, Muscle and Synapse". McGraw-Hill, New York.

Kendall, M. G. (1948). "The Advanced Theory of Statistics", Vol. 2. C. Griffin, London.

Khinchin, A. I. (1949). "Mathematical Foundations of Statistical Mechanics". Dover Publications, New York.

Khinchin, A. I. (1957). "Mathematical Foundations of Information Theory". Dover Publications, New York.

Khinchin, A. I. (1960). "Mathematical Methods in the Theory of Queueing". Griffin, London.

Kleene, S. C. (1956). Representation of events in nerve nets and finite automata. *In* "Automata Studies" (Eds C. E. Shannon and J. McCarthy), pp. 3–41. Princeton University Press, Princeton.

Kravitz, E. A. (1967). Acetylcholine, γ-aminobutyric acid, and glutamic acid: physiological and chemical studies related to their roles as neurotransmitter agents. *In* "The Neurosciences" (Eds G. C. Quarton, T. Melnechuk and F. O. Schmitt), pp. 433–444. The Rockefeller University Press, New York.

Kruskal, M. D. (1954). The expected number of components under a random mapping function. *Amer. math. Mon.* **61,** 392–397.

Kufler, S. W., Fitzhugh, R. and Barlow, H. B. (1957). Maintained activity in the cat's retina in light and darkness. *J. gen. Physiol.* **40,** 683–702.

Kuiper, J. W. and Leutscher-Hazelhoff, J. T. (1965). High-precision repetitive firing in the insect optic lobe and a hypothesis for its function in object location. *Nature, Lond.* **207,** 1158–1160.

Kuno, M. (1964). Quantal components of excitatory postsynaptic potentials in spinal motoneurones. *J. Physiol.* **175,** 81–99.

Lashley, K. S. (1929). "Brain Mechanisms and Intelligence". Reprinted (1963). Dover Publications, New York.

Lettvin, J. Y., Maturana, H. R., McCulloch, W. S. and Pitts, W. (1959). What the frog's eye tells the frog's brain. *Proc. Inst. Radio Engrs* **47,** 1940–1951.

Levick, W. R. and Williams, W. O. (1964). Maintained activity of lateral geniculate neurones in darkness. *J. Physiol., Lond.* **170,** 582–597.

Longuet-Higgins, H. C. (1968). The non-local storage of temporal information. *Proc. R. Soc.* B. **171,** 327–334.

Lotka, A. J. (1920). Undamped oscillations derived from the law of mass action. *J. Am. chem. Soc.* **42,** 1595–1599.

Lotka, A. J. (1925). "Elements of Physical Biology". Williams and Wilkins, Baltimore. Reprinted as "Elements of Mathematical Biology" by Dover Books, New York.

MacKay, D. K. and McCulloch, W. S. (1952). The limiting information capacity of a neuronal link. *Bull. math. Biophys.* **14,** 127–135.

Mahler, H. R. and Cordes, E. H. (1968). "Basic Biological Chemistry". Harper and Row, New York.

Marin-Padilla, M. (1968). Cortical axo-spinodendritic synapses in man: a Golgi study. *Brain Research, Osaka* **8,** 196–200.

Mark, R. F. (1969). Matching muscles and motoneurones. A review of some experiments on motor nerve regeneration. *Brain Research, Osaka* **14,** 245–254.

Martin, A. R. (1955). A further study of the statistical composition of the end-plate potential. *J. Physiol., Lond.* **130,** 114–122.

Martin, A. R. (1966). Quantal nature of synaptic transmission. *Physiol. Rev.* **46,** 51–66.

McCulloch, W. S. and Pitts, W. H. (1943). A logical calculus of ideas immanent in nervous activity. *Bull. math. Biophys.* **5,** 115–133.

McLennan, H. (1963). "Synaptic Transmission". W. B. Saunders, Philadelphia.

Minorsky, N. (1958). *In* "Dynamics and Non-linear Mechanics" by E. Leimanis and N. Minorsky. John Wiley, New York.

Monod, J. and Jacob, F. (1961). General conclusions: teleonomic mechanisms in cellular metabolism, growth and differentiation. *Cold Spring Harb. Symp. quant. Biol.* **26,** 389–401.

Nastuk, W. L. (1963, 1964). "Physical Techniques in Biological Research", Vols. 5 and 6: "Electrophysiological Methods", Academic Press, New York and London.

Oppenheimer, S. L. (1966). "Semiconductor Logic and Switching Circuits". C. E. Merrill Books, Columbus, Ohio.

Pakkenberg, H. (1966). The number of nerve cells in the cerebral cortex of man. *J. comp. Neurol.* **128,** 17–20.

Penfield, W. (1958). "The Excitable Cortex in Conscious Man". University Press, Liverpool.

Penfield, W. and Jasper, H. (1954). "Epilepsy and the Functional Anatomy of the Human Brain". J. and A. Churchill, London.

Pettigrew, J. D., Nikara, T. and Bishop, P. O. (1968). Responses to moving slits by single units in cat striate cortex. *Expl Brain. Res.* **6,** 373–390.

Pierce, J. R. (1961). "Symbols, Signals and Noise". Harper and Brothers, New York.

Pierce, W. (1967). "Failure-Tolerant Computer Design". Academic Press, New York and London.

Pirenne, M. H. (1967). "Vision and the Eye". Science Paperbacks, London.

Pitts, W. and McCulloch, W. S. (1947). How we know universals: the perception of auditory and visual forms. *Bull. math. Biophys.* **9,** 127–147.

Planck, M. (1890a). Ueber die Erregung von Electricität und Wärme in Electrolyten. *Ann. phys. Chem.* **39,** 161–186.

Planck, M. (1890b). Ueber die Potentialdifferenz zwischen zwei verdünnten Lösungen binärer Electrolyte. *Ann. phys. Chem.* **40,** 561–576.

Porter, B. (1967). "Stability Criteria for Linear Dynamical Systems". Oliver and Boyd, Edinburgh and London.

Pribram, K. H. (1966). Some dimensions of remembering: steps toward a neuropsychological model of memory. *In* "Macromolecules and Behavior" (Ed. J. Gaito), pp. 165–186. Appleton-Century Crofts, New York.

Pringle, J. W. S. (1951). On the parallel between learning and evolution. *Behaviour* **3,** 174–215.

Pye, K. and Chance, B. (1966). Sustained sinusoidal oscillations of reduced pyridine nucleotide in a cell-free extract of saccharomyces carlsbergensis. *Proc. natn. Acad. Sci. U.S.A.* **55,** 888–894.

Quastler, H. (1956). Studies of human channel capacity. *In* "Information Theory" (Ed. C. Cherry), pp. 361–371. Academic Press, New York and London.

Quastler, H. (1965). "Information Theory in Psychology." The Free Press, Illinois.

Racah, G. (1942). Theory of complex spectra II. *Phys. Rev.* **62,** 438–462.

Ransom, S. W. and Clark, S. L. (1961). "The Anatomy of the Nervous System". W. B. Saunders, Philadelphia.

Rapoport, A. (1952). Ignition phenomena in random nets. *Bull. math. Biophys.* **14,** 35–44.

Rapoport, A. (1955). Application of information networks to a theory of vision. *Bull. math. Biophys.* **17,** 15–33.

Rashevsky, N. (1945). A suggestion for another statistical interpretation of the fundamental equations of the mathematical biophysics of the central nervous system. *Bull. math. Biophys.* **7,** 223–226.

Richter, D. (1966), Editor. "Aspects of Learning and Memory". William Heinemann, London.

Riesen, A. H. (1947). The development of visual perception in man and chimpanzee. *Science, N.Y.* **106,** 107–108.

Riordan, J. (1962). Enumeration of linear graphs for mappings of finite sets. *Ann. math. Stat.* **33,** 178–185.

Roberts, E. (1966). The synapse as a biochemical self-organizing micro-cybernetic unit. *Brain Research, Osaka* **1,** 117–166.

Roberts, M. B. V. (1962). The rapid response of *Myxicola infundibulum* (Grübe). *J. mar. biol. Ass. U.K.* **42,** 527–539.

Roeder, K. D. (1963). "Nerve Cells and Insect Behavior". Harvard University Press, Cambridge, Mass.

Rosenblatt, F. (1967). Recent work on theoretical models of biological memory. *Comput. Inform. Sci.* **2,** 33–56.

Ruch, T. C., Patton, H. D., Woodbury, J. W. and Towe, A. L. (1964). "Neurophysiology". W. B. Saunders, Philadelphia.

Russell, I. S. (1966). Animal learning and memory. *In* "Aspects of Learning and Memory" (Ed. D. Richter), p. 164. William Heinemann, London.

Russell, W. R. (1948). Studies in amnesia. *Edinb. med. J.* **50,** 92–99.
Russell, W. R. (1959). "Brain Memory Learning". Oxford University Press, London.
Rutherford, D. E. (1954). "Vector Methods". Oliver and Boyd, London.
Salmoiraghi, G. C. and Von Baumgarten, R. (1961). Intracellular potentials from respiratory neurones in brain-stem of cat and mechanism of rhythmic respiration. *J. Neurophysiol.* **24,** 203–218.
Schlögl, R. (1954). Elektrodiffusion in freier Lösung und geladenen Membranen. *Z. phys. Chem., Frankf. Ausg.* **1,** 305–339.
Scott, J. P. (1962). Critical periods in behavioral development. *Science, N.Y.* **138,** 949–958.
Shannon, C. E. and Weaver, W. (1949). "The Mathematical Theory of Communication". University of Illinois Press, Urbana, Illinois.
Sherrington, C. S. (1940). "Man on His Nature". Cambridge University Press, London.
Shimbel, A. (1950). Contributions to the mathematical biophysics of the central nervous system with special reference to learning. *Bull. math. Biophys.* **12,** 241–275.
Sholl, D. A. (1956). "The Organization of the Cerebral Cortex". Methuen, London.
Sperry, R. W. (1947). Cerebral regulation of motor coordination in monkeys following multiple transection of sensorimotor cortex. *J. Neurophysiol.* **10,** 275–294.
Sperry, R. W. (1961). Cerebral organization and behavior. *Science, N.Y.* **133,** 1749–1757.
Sperry, R. W. and Miner, N. (1955). Pattern perception following insertion of mica plates into visual cortex. *J. comp. Physiol. Psychol.* **48,** 463–469.
Stevens, C. F. (1964). (Letter to the Editor). *Biophys. J.* **4,** 417–419.
Stoker (1950). "Non-linear Vibrations". Interscience Publishers, New York.
Sumitomo, I., Ide, K. and Iwama, K. (1969). Conduction velocity of rat optic nerve fibers. *Brain Research, Osaka* **12,** 261–264.
Tapper, D. N. and Mann, M. D. (1968). Single presynaptic impulse evokes postsynaptic discharge. *Brain Research, Osaka* **11,** 688–690.
Tasaki, I. (1959). Conduction of the nerve impulse. *In* "Handbook of Physiology: Neurophysiology" (Eds J. Field, H. W. Magoun and V. E. Hall), Vol. 1, pp. 75–121. American Physiological Society, Washington, D.C.
Tasaki, I. (1967). "Nerve Excitation. A Macromolecular Approach". C. C. Thomas, Springfield, Ill.
Thorpe, W. H. (1961). Bird Song". Cambridge University Press, London.
Tolman, R. C. (1938). "The Principles of Statistical Mechanics". Oxford University Press, London.
Tower, D. B. (1954). Structural and functional organization of mammalian cerebral cortex: the correlation of neurone density with brain size. *J. comp. Neurol.* **101,** 19–46.
Truex, R. C. and Carpenter, M. B. (1969). "Human Neuroanatomy". Williams and Wilkins, Baltimore.
Turing, A. M. (1950). Computing machinery and intelligence. Reprinted in "Minds and Machines" (Ed. A. R. Anderson), pp. 4–30. Prentice-Hall, New York.
Uspensky, J. V. (1948). "Theory of Equations". McGraw-Hill, New York.
Vajda, S. (1966). "An Introduction to Linear Programming and the Theory of Games". Science Paperbacks, London.

Vallecalle, E. and Svaetichin, G. (1961). The retina as model for the functional organization of the nervous system. In "Neurophysiologie und Psychophysik des visuellen Systems" (Eds R. Jung and H. Kornhuber), pp. 489–492. Springer-Verlag, Berlin.

Volterra, V. (1931). "La Théorie Mathématique de la lutte pour la vie". Gauthiers-Villars, Paris.

Von Neumann, J. (1956). Probabilistic logics and the synthesis of reliable organisms from unreliable components. *In* "Automata Studies" (Eds C. E. Shannon and J. McCarthy), pp. 43–98. Princeton University Press, Princeton.

Von Neumann, J. (1958). "The Computer and the Brain". Yale University Press, New Haven.

Von Neumann, J. and Morgenstern, O. (1955). "Theory of Games and Economic Behavior". Princeton University Press, Princeton.

Wachtel, H. and Kandel, E. R. (1967). A direct synaptic connection mediating both excitation and inhibition. *Science, N.Y.* **158,** 1206–1208.

Wall, P. D., Lettvin, J. Y., McCulloch, W. S. and Pitts, W. H. (1956). Factors limiting the maximum impulse transmitting ability of an afferent system of nerve fibres. *In* "Information Theory", third London Symposium (Ed. C. Cherry), pp. 329–344. Academic Press, New York and London.

Watson, J. D. (1965). "Molecular Biology of the Gene". W. A. Benjamin, New York.

Weisskrantz, L. (1970). A long-term view of short-term memory in psychology. *In* "Short-term Changes in Neural Activity and Behaviour" (Eds R. A. Hinde and G. Horn), pp. 63–74. Cambridge University Press, London.

Werner, G. and Mountcastle, V. B. (1963). The variability of central neural activity in a sensory system, and its implications for the central reflection of sensory events. *J. Neurophysiol.* **26,** 958–977.

Whittaker, E. T. and Watson, G. N. (1952). "A Course of Modern Analysis". Cambridge University Press, London.

Wiener, N. (1961). "Cybernetics". John Wiley, New York.

Wiersma, C. A. G. (1961). Reflexes and the central nervous system. *In* "The Physiology of Crustacea" (Ed. T. H. Waterman), Vol. 2, pp. 241–279. Academic Press, New York and London.

Willshaw, D. J., Buneman, O. P. and Longuet-Higgins, H. C. (1969). Non-holographic associative memory. *Nature, Lond.* **222,** 960–962.

Wooldridge, D. (1963). "The Machinery of the Brain". McGraw-Hill, New York.

Wyburn, G. M. (1960). "The Nervous System". Academic Press, London and New York.

Young, J. Z. (1964). "A Model of the Brain". Clarendon Press, Oxford.

Young, J. Z. (1966). "The Memory System of the Brain". Oxford University Press, London.

Zeeman, E. C. (1962). The topology of the brain and visual perception. *In* "Topology of 3-manifolds and Related Topics" (Ed. M. K. Fort), pp. 240–256. Prentice-Hall, New Jersey.

Subject Index

The index is largely for the purpose of obtaining definitions of frequently used words or symbols, although it does contain other items as well.

Symbols used in the text

Symbols	*On page*	*Symbols*	*On page*
H	94	t_o	29
m	23	δ	35
n_e	30	ε	35
n_i	30	η	35
N_e	30	θ	29
N_i	30	τ	29, 34
p	23	$\tau_1, \tau_2, \tau_3, \tau_4$	34
p_o	72	Φ	30
$P(n, \theta, p)$	68	$\rightarrow$	31
q	23	$\leftrightarrow$	31
R	35